CLIMATE CHANGE AND ITS CAUSES, EFFECTS AND PREDICTION

CLIMATE CHANGE

DEPARTMENT OF DEFENSE INFRASTRUCTURE ADAPTATION EFFORTS

Climate Change and its Causes, Effects and Prediction

Additional books in this series can be found on Nova's website under the Series tab.

Additional e-books in this series can be found on Nova's website under the e-book tab.

CLIMATE CHANGE AND ITS CAUSES, EFFECTS AND PREDICTION

CLIMATE CHANGE

DEPARTMENT OF DEFENSE INFRASTRUCTURE ADAPTATION EFFORTS

EZRA STACK
EDITOR

NOVINKA

New York

For permission to use material from this book please contact us:
Telephone 631-231-7269; Fax 631-231-8175
Web Site: http://www.novapublishers.com

Library of Congress Cataloging-in-Publication Data

ISBN: 978-1-63463-383-3

Published by Nova Science Publishers, Inc. † New York

CONTENTS

PREFACE

In its Fiscal Year 2012 Climate Change Adaptation Roadmap, the Department of Defense (DOD) identified climate change phenomena such as rising temperatures and sea levels as potentially impacting its infrastructure, and officials at sites GAO visited or contacted noted actual impacts they had observed. This book describes potential impacts identified by DOD that may affect its infrastructure; evaluates DOD's efforts to conduct vulnerability assessments; assesses how DOD is accounting for climate change impacts in certain planning efforts; and evaluates the extent to which DOD incorporates adaptation into its infrastructure-investment efforts.

Chapter 1 – According to DOD, its U.S. infrastructure is vulnerable to the potential impacts of climate change. These could affect DOD's readiness and fiscal exposure, and DOD has begun to plan for adaptation actions designed to improve infrastructure resilience. GAO was asked to assess DOD's actions to adapt its U.S. infrastructure to the challenges of climate change. This report (1) describes potential impacts identified by DOD that may affect its infrastructure; (2) evaluates DOD's efforts to conduct vulnerability assessments; (3) assesses how DOD is accounting for climate change impacts in certain planning efforts; and (4) evaluates the extent to which DOD incorporates adaptation into its infrastructure-investment efforts.GAO reviewed DOD documents on climate change, infrastructure planning, and funding processes; interviewed cognizant officials; and visited or contacted a nongeneralizable sample of 15 sites in various regions and from each military department.

Chapter 2 – The Department of Defense (DoD) Climate Change Adaptation Roadmap (CCAR) fulfills a requirement of Executive Order 13514, *Federal Leadership in Environmental, Energy and Economic Performance.* All Federal Departments and Agencies should evaluate climate change risks and vulnerabilities to manage both the short- and long-term effect of climate change on the agency's mission and operations, and include an adaptation planning document as an appendix to its annual Strategic Sustainability Performance Plan.

In: Climate Change
Editor: Ezra Stack

ISBN: 978-1-63463-383-3

Chapter 1

CLIMATE CHANGE ADAPTATION: DOD CAN IMPROVE INFRASTRUCTURE PLANNING AND PROCESSES TO BETTER ACCOUNT FOR POTENTIAL IMPACTS*

United States Government Accountability Office

WHY GAO DID THIS STUDY

According to DOD, its U.S. infrastructure is vulnerable to the potential impacts of climate change. These could affect DOD's readiness and fiscal exposure, and DOD has begun to plan for adaptation actions designed to improve infrastructure resilience. GAO was asked to assess DOD's actions to adapt its U.S. infrastructure to the challenges of climate change. This report (1) describes potential impacts identified by DOD that may affect its infrastructure; (2) evaluates DOD's efforts to conduct vulnerability assessments; (3) assesses how DOD is accounting for climate change impacts in certain planning efforts; and (4) evaluates the extent to which DOD incorporates adaptation into its infrastructure-investment efforts.GAO reviewed DOD documents on climate change, infrastructure planning, and funding processes; interviewed cognizant officials; and visited or contacted a

* This is an edited, reformatted and augmented version of a United States Government Accountability Office publication, No. GAO-14-446, dated May 2014.

nongeneralizable sample of 15 sites in various regions and from each military department.

WHAT GAO RECOMMENDS

GAO recommends that DOD develop a plan and milestones for completing climate change vulnerability assessments of installations; provide further information to installation planners, clarifying actions that should be taken to account for climate change in planning documents; and clarify the processes used to compare military construction projects for funding, to include consideration of potential climate change impacts. DOD concurred with GAO's recommendations and explained how they will be implemented.

WHAT GAO FOUND

In its Fiscal Year 2012 Climate Change Adaptation Roadmap, the Department of Defense (DOD) identified climate change phenomena such as rising temperatures and sea levels as potentially impacting its infrastructure, and officials at sites GAO visited or contacted noted actual impacts they had observed. For example, according to DOD officials, the combination of thawing permafrost, decreasing sea ice, and rising sea levels on the Alaskan coast has increased coastal erosion at several Air Force radar early warning and communication installations. Impacts on DOD's infrastructure from this erosion have included damaged roads, seawalls, and runways. In addition, officials on a Navy installation told GAO that sea level rise and resulting storm surge are the two largest threats to their waterfront infrastructure. For instance, they are concerned about possible storm surge during work on a submarine that will be cut in half while sitting in a dry dock. Officials explained that if salt water floods the submarine's systems, it could result in severe damage.

DOD has begun to assess installations' vulnerability to potential climate change impacts and directed its planners to incorporate consideration of climate change into certain installation planning efforts. Further, it is a DOD strategic goal to consider sustainability, including climate change adaptation, in its facility investment decisions. However, GAO identified some limitations with these efforts. Specifically:

- DOD has begun collecting data on historic and potential future vulnerabilities from coastal locations (installations and associated sites) and is developing regional sea-level rise scenarios for 704 coastal locations to be used following the collection of these data. However, it has not yet developed a plan or milestones for completing these tasks, including when it expects to finish data collection on a total of 7,591 locations worldwide. Without a plan, including interim milestones to gauge progress, DOD may not finish its assessments in a timely and complete manner.
- DOD guidance requires that both installation master planning and natural resources planning account for certain potential impacts of climate change, but the implementation of these requirements across the department varies. Installation planners said that they lack key definitions and updated guidance on construction and renovation going beyond current building codes to account for climate change. Without additional information, installation planners will be unlikely to consistently account for climate change impacts in their Master Plans and Integrated Natural Resources Management Plans.
- Installation officials rarely propose climate change adaptation projects because the services' processes for approving and funding military construction projects do not include climate change adaptation in the criteria used to rank potential projects. As a result, installation planners may believe that climate change adaptation projects are unlikely to successfully compete with other military construction projects for funding. Without clarification of these processes, DOD may face challenges in meeting its strategic goals and the services may miss opportunities to make their facilities more resilient to the potential impacts of climate change.

ABBREVIATIONS

Assessment Group	Assessment Guidance Sub-Working Group
DOD	Department of Defense
Roadmap	Fiscal Year 2012 Climate Change Adaptation Roadmap
Scenario Group	Coastal Assessment Regional Scenario Working

May 30, 2014

Congressional Requesters:

The Department of Defense (DOD) manages a global real-estate portfolio that includes over 555,000 facilities and 28 million acres of land with a replacement value of close to $850 billion. Within the United States, the department's extensive infrastructure of bases and training ranges,[1] which is critical to maintaining military readiness, extends across all regions, as well as Alaska, and Hawaii. This infrastructure is vulnerable to the potential impacts of climate change.[2] Scientific projections and observations indicate that these potential impacts include increasing temperatures, rising sea levels, melting permafrost and other gradual changes, as well as the potential for increases in the frequency and severity of extreme events. We have previously reported that while it is not possible to link any individual weather event to climate change, these events provide insight into the potential climate-related vulnerabilities the United States faces.[3] In the Third National Climate Assessment,[4] the U.S. Global Change Research Program[5] states that all regions of the country will face climate change impacts that affect natural ecosystems as well as the nation's infrastructure. In addition, infrastructure across the United States is being adversely affected already by phenomena associated with climate change, including sea level rise, storm surge, heavy downpours, and extreme heat according to the assessment. The assessment further notes that proactively preparing for climate change can reduce impacts, while also facilitating a more-rapid and efficient response to changes as they happen.

GAO, DOD, and others have cited the risks posed by climate change and the need to engage in adaptation activities. In February 2013, we added the federal government's efforts to manage its fiscal exposure to the effects of climate change to our High Risk list, 6 noting that the federal government's role as a property owner exposes it to significant fiscal risk. We also noted that the government currently lacks a shared understanding of strategic priorities and adequate interagency coordination to adapt to a changing climate.

DOD has noted the challenges that the impacts of climate change may pose for the department, including to its facilities. In the 2010 Quadrennial Defense Review, 7 DOD stated that many coastal installations were already facing elevated levels of risk from rising sea levels and that the department needs to complete a comprehensive assessment of all its installations to determine mission impacts and to adapt as required. In its fiscal year 2012

Climate Change Adaptation Roadmap (Roadmap),[8] DOD noted that the department's ability to adapt to climate change will be a significant factor in its ability to fulfill its missions in the future, and that climate-related effects already are being observed at DOD installations throughout the United States. Furthermore, the Roadmap states that including climate change and climate variability considerations in the department's planning processes will enhance operational and infrastructure resilience. In its 2014 Quadrennial Defense Review, DOD reaffirmed its 2010 assessment of the risks that climate change impacts pose to the department's infrastructure and its finding that DOD needs to adapt as required. Given its extensive infrastructure, DOD has begun to plan for climate change impacts on its infrastructure.

We were asked to assess DOD's progress in taking action to adapt its U.S. infrastructure to the challenges of climate change. 9 In this report, we (1) describe the climate change phenomena and associated impacts and mission vulnerabilities that DOD has identified; (2) evaluate the extent to which DOD has conducted, or plans to conduct, vulnerability assessments of climate change impacts on installation infrastructure; (3) assess how DOD is accounting for climate change impacts in installations' selected planning efforts for existing and future infrastructure; and (4) evaluate how DOD has incorporated adaptation to climate change impacts into infrastructure investment efforts.

To describe DOD's efforts to identify climate change phenomena and associated impacts and vulnerabilities on its infrastructure, we reviewed documents such as the 2010 and 2014 Quadrennial Defense Reviews, DOD's fiscal year 2012 Strategic Sustainability Performance Plan, and the Roadmap. We also reviewed ongoing studies from DOD's Strategic Environmental Research and Development Program on potential climate change impacts at DOD locations, and met with officials from the Office of the Secretary of Defense and the military department headquarters. To gain insight into examples of climate change phenomena that have or may impact installation infrastructure in the future and that may lead to mission vulnerabilities, we visited or contacted officials at a nongeneralizable sample of installations, learned about or observed impacts on infrastructure, and collected key documentation such as after-action reports describing impacts and mission vulnerabilities of climate change phenomena. The nongeneralizable sample of installations that we selected consisted of 15 locations, and included a variety of geographic locations and each of the military departments. The selected locations had been—or were expected by DOD to be—impacted by climate change. Although the information we collected was not representative of all installations, we determined that the selection of these installations would provide valuable insights for our review.

To determine the extent to which DOD has assessed or plans to assess the vulnerability of its facilities to the potential impacts of climate change, we reviewed DOD's 2010 and 2014 Quadrennial Defense Review and Roadmap to determine DOD's goals for conducting vulnerability assessments. To determine the extent to which the department has completed such assessments, we reviewed DOD documentation that discusses vulnerability assessments and interviewed officials in the Office of the Secretary of Defense and military departments to learn about DOD's assessment efforts. In addition, we observed meetings of DOD's key organizations involved in these assessment efforts. We also reviewed project management practices of high-performing organizations, comparing DOD's assessment efforts to these practices in order to determine the extent to which DOD has incorporated them in its assessment efforts.

To determine the extent to which DOD is accounting for climate change impacts in selected installation planning efforts for infrastructure, we identified the requirements for doing so by reviewing DOD's Unified Facilities Criteria for Installation Master Planning[10] and DOD Instruction 4715.03[11] that set these requirements for Installation Master Planning (Master Plans) and Integrated Natural Resources Management Plans, respectively. We also met with officials from the selected installations who were responsible for the Master Plan and Integrated Natural Resources Management Plans to learn about how they incorporated the requirements into their processes. We also compared DOD's requirements with Master Plans and Integrated Natural Resources Management Plans from the nongeneralizable sample of 15 installations we visited or contacted. In addition, we reviewed *The Standards for Internal Control in the Federal Government* related to information and communications,[12] and compared those standards to DOD's practices related to Installation Master Planning and Integrated Natural Resources Management Plans.

To determine the extent to which DOD accounts for climate change impacts when making infrastructure project funding decisions, we reviewed DOD's fiscal year 2012 Strategic Sustainability Performance Plan and Roadmap to determine DOD's policies and goals for integrating climate change adaptation actions into facility investment decision processes. We also interviewed DOD officials at the military department headquarters level and at the selected installations we visited or contacted to determine the extent to which the military departments have implemented climate change adaptation efforts at the installation level. In addition, we reviewed examples of military department processes used for approval and funding of potential military construction projects. Finally, we compared the extent to which the processes incorporate climate change adaptation with DOD's policies and goals on integrating adaptation actions into facility investment decisions.

We conducted this performance audit from May 2013 to May 2014 in accordance with generally accepted government auditing standards. Those standards require that we plan and perform the audit to obtain sufficient, appropriate evidence to provide a reasonable basis for our findings and conclusions based on our audit objectives. We believe that the evidence obtained provides a reasonable basis for our findings and conclusions based on our audit objectives. Further details on our scope and methodology can be found in appendix I.

BACKGROUND

Adaptation and Climate Change

We have previously reported[13] that changes in the Earth's climate attributable to increased concentrations of greenhouse gases may have significant environmental and economic impacts in the United States.[14] Proposed responses to climate change include reducing greenhouse gas emissions through regulation, promoting low-emissions technologies, and adapting to the possible impacts by planning and improving protective infrastructure. In recent years, climate change adaptation has begun to receive more attention because—as we and others have reported—the greenhouse gases already in the atmosphere are expected to continue altering the climate system into the future, regardless of efforts to control emissions.[15] Climate change adaptation differs from mitigation, which is focused on reducing emissions. In this report, we focus on adaptation efforts. For a summary of our previous work on U.S. government climate change adaptation efforts and related recommendations for improvement of these efforts, see appendix II.

Current Understanding of Global Climate Change Is Based on Both Observations and Computer Model Projections

According to the Third National Climate Assessment, scientists' current understanding of how the global climate is changing is based on both observations and projections reached through the use of computer simulations, using global climate models, that account for a variety of independent factors.[16] The models used to project changes are becoming more sophisticated as scientists incorporate further variables into them.

Generally, the models' projections indicate that higher concentrations of greenhouse gasses will result in greater climate change, thus increasing the degree to which the nation, including its infrastructure, is exposed to risk. Nonetheless, uncertainty remains about projections of future changes and most projections indicate ranges of change rather than specific figures. The effects of increases in atmospheric concentrations of greenhouse gases and temperature are expected to have varying impacts in the United States (see table 1).

Table 1. Current and Projected Impacts of Climate Change in the United States

Category	Current and projected impacts
Temperature	U.S. average temperature has risen fewer than 2 degrees Fahrenheit over the past 50 years. It is projected to rise more in the future—how much more depends primarily on the amount of heat-trapping gases emitted globally and how sensitive the climate is to those emissions.
Precipitation	Precipitation in the U.S. has increased an average of about 5 percent over the past 50 years. Projections of future precipitation generally indicate that northern areas will become wetter and southern areas, particularly in the West, will become drier. The amount of rain falling in the heaviest downpours has increased approximately 20 percent on average in the past century, and this trend isvery likely to continue, with the largest increases in the wettest places.
Extreme weather events	Some types of extreme weather events, such as heat waves and regional droughts, have become more frequent and intense during the past 40 to 50 years.

Source: Adapted from U.S. Global Change Research Program and National Research Council data.

Executive Orders and DOD Strategic Documents Direct DOD to Plan for and Adapt to Climate Change

At the government-wide level, the President has released two executive orders directing agencies to plan for and adapt to certain impacts of climate change. In 2009, the President signed Executive Order 13514,[17] calling for federal agencies to take a number of actions related to climate change. In addition, the President's Council on Environmental Quality issued implementing instructions directing federal agencies to, among other things, establish an agency climate change adaptation policy, complete a high-level analysis of agency vulnerability to climate change, and deliver to the council and the Office of Management and Budget a climate adaptation plan for implementation in fiscal year 2013.[18] In response to these implementing instructions, DOD developed the high-level Roadmap that it included as an appendix to its fiscal year 2012 Strategic Sustainability Performance Plan. We discuss both this plan and the Roadmap in greater detail below.

On November 1, 2013, the President released Executive Order 13653, 19 which directs federal agencies to develop or continue to develop, implement, and update comprehensive agency climate change adaptation plans that address a number of efforts. The order requires agency plans to integrate consideration of climate change into their operations and overall mission objectives. Also, the order stipulates that the plans should include, among other things, a description of how agencies will consider the need to improve climate adaptation, with respect to, among other things, real-property investments, through actions such as updating agency policies for leasing, building upgrades, relocation of existing facilities and equipment, and construction of new facilities.

As we noted previously, DOD has—in several strategic documents—cited the negative effects that climate change could have on its infrastructure and operations. In its 2010 Quadrennial Defense Review, DOD stated that the department's operational readiness hinges on continued access to land, air, and sea training and test space and that, consequently, DOD must complete an assessment of all its installations to determine potential climate change impacts on its missions and adapt as required. Subsequently, in the 2011 National Military Strategy, DOD's characterization of the strategic environment included climate change as a potentially serious impact. Also, in its fiscal year 2012 Strategic Sustainability Performance Plan, DOD states that climate change can directly impact military installations, infrastructure, and operations by limiting the availability and quality of training ranges and other lands needed for operations, and by increasing impacts on infrastructure such as flood and fire hazards and vulnerability of utilities. In addition, in its 2013 Arctic Strategy, DOD recognized that climate change will increase access and activity in the region, potentially altering the security environment in which it operates. Further, in its 2014 Arctic Roadmap, the Navy acknowledges the role that climate change plays in several national security arenas. For example, the Navy states that the Arctic is warming faster than the rest of the world—and as a result—significant retreat of sea ice will accelerate throughout this century, causing previously unreachable areas to be increasingly open for maritime use.[20] Finally, in its 2014 Quadrennial Defense Review, DOD states that the impacts of climate change may undermine the capacity of the department's domestic installations to support training activities.

Several DOD Organizations Play a Role in the Department's Climate Change Adaptation Efforts

DOD and each of the military departments have given certain organizations key roles in the department's climate change adaptation efforts. DOD has designated the Under Secretary of Defense for Acquisition, Technology and Logistics as the department's Senior Sustainability Officer responsible for ensuring the effective and successful implementation of the department's strategic sustainability initiatives described in the fiscal year 2012 Strategic Sustainability Performance Plan, which includes climate change adaptation. Also, DOD has established a governance structure to facilitate meeting the department's sustainability goals. Reporting to the Senior Sustainability Officer, there is a Senior Sustainability Council that consists of representatives from the military departments and other DOD organizations. A set of committees and working groups has been established with the goal of executing the goals of the fiscal year 2012 Strategic Sustainability Performance Plan. The committees and working groups cover a wide range of sustainability topics, including transportation and fuels, solid waste and recycling, and climate change adaptation for DOD infrastructure. We discuss a number of these working groups in more detail later in the report.

In addition, each military department has both designated a sustainability officer with the responsibility for ensuring the effective and successful implementation of the department's strategic sustainability initiatives—as outlined in the fiscal year 2012 Strategic Sustainability Performance Plan—

and appointed representatives to serve on the committees and working groups described above. Also, certain organizations within the Office of the Secretary of Defense and each military department play a key role in the development of climate change adaptation policy. These organizations include

- the Office of the Deputy Under Secretary of Defense for Installations and Environment;
- the Office of the Assistant Secretary of the Army for Installations, Energy and Environment;
- the Office of the Assistant Secretary of the Air Force for Installations, Environment, and Logistics; and
- the Office of the Assistant Secretary of the Navy for Energy, Installations, and Environment.

DOD also works with other federal agencies such as the Environmental Protection Agency and National Oceanic and Atmospheric Administration on issues related to climate change adaptation efforts. For example, the Strategic Environmental Research and Development Program is a partnership between DOD, the Environmental Protection Agency, and the Department of Energy to address environmental issues at DOD and at the Department of Energy through investing in a broad spectrum of basic and applied research.

DOD Has Identified Climate Change Phenomena, Potential Impacts, and Potential Mission Vulnerabilities, and DOD Officials Have Observed Them at Selected U.S. Installations

In its Roadmap, DOD describes five categories of climate change phenomena and notes that climate-related effects are already being observed at DOD installations throughout the United States. DOD further indicates that these phenomena could translate into potential climate change impacts and vulnerabilities that could adversely affect infrastructure across all of DOD's facilities. Although officials at the locations we visited or contacted generally agreed that individual weather events—such as a single storm—cannot be tied directly to changes in the climate, these officials noted that DOD facilities and infrastructure are vulnerable to climate change phenomena and recognized that climate change may make these types of phenomena more frequent or severe.

DOD's Roadmap Identifies Five Categories of Climate Change Phenomena, Potential Impacts, and Potential Mission Vulnerabilities

DOD describes five climate change phenomena in its Roadmap: (1) rising temperatures; (2) changes in precipitation patterns; (3) increasing storm frequency and intensity (coastal and inland); (4) rising sea levels and associated storm surge; and (5) changes in ocean temperature, circulation, salinity, and acidity. Further, in the Roadmap, DOD associates each phenomenon with potential impacts or vulnerabilities to its infrastructure. For instance, DOD associates rising temperatures with potential climate change impacts such as thawing permafrost and wildfire risk. According to the Roadmap, this may result in mission vulnerabilities such as reduced military vehicle access and potential loss of cold weather training venues. In addition, DOD links changes in precipitation patterns with potential climate change impacts such as changes in the number of consecutive days of high or low precipitation as well as increases in the extent and duration of droughts, with an associated increase in the risk of wildfire. According to the Roadmap, this may result in mission vulnerabilities such as reduced live-fire training due to drought and increased wildfire risk, reduced water availability, and increased flood control or erosion prevention measures. Further, DOD associates increasing storm frequency and intensity (coastal and inland) with potential climate change impacts such as flooding, soil and vegetation loss, and wind damage. According to the Roadmap, this may result in mission vulnerabilities such as temporary or prolonged disruption of military operations or test and training activities, increased maintenance costs, and damage to coastal infrastructure. DOD associates rising sea levels and associated storm surge with potential climate change impacts such as damage to physical infrastructure and salt water intrusion.

According to the Roadmap, these impacts may result in mission vulnerabilities such as increased cost of infrastructure modification, for instance raising piers, impacts to future land availability, and siting of new construction. DOD associates changes in ocean temperature, circulation, salinity, and acidity with potential climate change impacts such as coral reef losses that may negatively impact the reef's ability to mitigate the effects of storm surge. According to the Roadmap, this may result in mission vulnerabilities such as increased cost of infrastructure reinforcement to withstand increased storm intensities and impacts to shore training and ranges.

Installation Officials Provided Examples of Potential Mission Vulnerabilities Associated with Impacts to Infrastructure

During our site visits, installation officials identified potential climate change phenomena experienced at their installations;[21] provided examples of potential impacts on installation infrastructure; discussed potential mission vulnerabilities, including potential impacts to military readiness; and provided the following examples of these impacts.

Rising Temperatures

At 8 out of 15 locations we visited or contacted,[22] officials stated that they had observed rising temperatures and associated potential impacts or mission vulnerabilities, as described by DOD in the Roadmap. For example, the combination of thawing permafrost, decreasing sea ice, and rising sea level on the Alaskan coast have led to an increase in coastal erosion at several Air Force radar early warning and communication installations. According to installation officials, this erosion has damaged roads, utility infrastructure, seawalls, and runways. For example, at one radar early warning installation, 40 feet of shoreline has been lost as a result of erosion and the erosion has damaged half of the runway.

As a result, only small planes or helicopters are able to land in this location, as opposed to larger planes that could land on the runway when it is fully functional. This means that access to the radar installation is limited. At another radar early warning installation, increased erosion has damaged a seawall, allowing increasingly large waves to damage the overrun area of a runway[23] (see figure 1).

According to installation officials, daily operations at these types of remote radar installations are at risk due to potential loss of runways, and such installations located close to the coastline could be at risk of radar failure if erosion of the coastline continues. Air Force headquarters officials noted that if one or more of these sites is not operational, there is a risk that the DOD early warning system will operate with diminished functionality. However, the system is designed to operate with the loss of one or more sites, as other sites can back up those lost.

Source: U.S. Air Force.
Note: Increased coastal erosion has allowed waves to damage the overrun area of the runway at an Air Force radar early warning installation.

Figure 1. Overrun area of Air Force runway impacted by erosion.

Also, officials at an Army installation told us that rising temperatures are impacting certain training activities because of thawing permafrost. They explained that in the summer of 2003, they experienced an unusually warm winter that caused permafrost to thaw and forced DOD to halt training for approximately 3 weeks because the ground was too soft to traverse. Officials stated that if temperatures continue to rise as projected, permafrost thawing could become more severe. This could further impact DOD training and may impact military readiness because DOD could not easily find another location to replicate the training offered in this area. Specifically, thawing permafrost has caused holes to open up in the ground of a drop zone.[24] These holes make certain areas of the drop zone unsafe for airborne training. As a result, this type of training may be limited. This is especially significant given that thawing permafrost occurs during the warmer months—when there is the heaviest demand for training—and this is the only drop zone in the training area. Further, according to officials, thawing permafrost may impact species migration patterns as a result of potential changes to wetlands. Officials told us that—depending on local topography and the makeup of local soil—the loss of permafrost can either make land drier or wetter.[25] These changes can degrade existing wetlands or create new ones. These officials explained that because of laws and regulations that govern DOD's management of wetlands and the species that inhabit them, the movement of wetlands and associated species to new areas may limit DOD's ability to train in certain sections of a training area. Further, officials told us that they have noticed an increase in freezing rain due to rising temperatures. In the past, colder temperatures typically produced snow as opposed to freezing rain. This rain has affected targets that now require additional maintenance. Specifically, ice buildup from freezing rain can lock targets' moving parts, breaking the targets or stopping them from properly functioning. Depending on the severity of the ice buildup, this may result in delays to training schedules.

Changes in Precipitation Patterns

At 9 out of 15 locations we visited or contacted, officials stated that they had observed changes in precipitation patterns and associated potential impacts. For example, officials at an installation in the Pacific told us that in 2008 they experienced 43 straight days of rain that resulted in mud slides and flooding that damaged base infrastructure, including base housing. By contrast, officials provided examples of impacts from reduced precipitation—such as drought and wildfire risk—and identified potential mission vulnerabilities—such as reduced live-fire training—as described by DOD in the Roadmap. For instance, officials also told us that for the last 3 to 4 years, there has been a continuous drought at another nearby installation, which has led to an increased number of wildfires in the area. As a result of a 2012 wildfire, officials were unable to access or move ammunition at that installation for 4 days. At a third installation in the same area, officials explained that these drought conditions—and the threat of wildfire—have limited the types of ammunition that can be used on certain training ranges (see fig. 2). As a result, units have had to spend extra time and money to travel to other installations to complete their required training.

Source: GAO.

Note: Drought conditions and an increased risk of wildfire limited the types of ammunition that units could use on this training range.

Figure 2. Training Range in the Pacific.

Further, at an Army training area, officials told us drought contributed to wildfires in 2013 and that units that had planned to train in the area experienced several mission impacts (see fig. 3). Officials told us that they had to postpone training from June 2013 to September or October 2013, and that the fires' smoke limited units' use of certain weapons systems in training. For example, there was no live-fire training allowed in one training area for 2 months. In addition, there was a decrease in the realism of training conducted. For instance, during an exercise that coincided with the fires, an official explained that aircraft could not deliver ordnance. The fires also affected landing conditions, forcing aircraft to save fuel, and thus reducing the amount of tactical training they were able to conduct.

Source: Alaska Fire Service.

Note: Drought conditions contributed to a 2013 fire that limited the use of certain weapons systems and training activities.

Figure 3. Wildfire on a training range in Alaska.

Increased Storm Frequency and Intensity

At 12 out of 15 locations we visited or contacted, officials stated that they had observed increasing storm frequency and intensity and associated potential impacts such as flooding, or mission vulnerabilities such as temporary or prolonged disruption of military operations, as described by DOD in the Roadmap. For example, officials at an installation located in the desert Southwest explained that flash flooding due to an intense rain event had made one of their emergency runways unusable. According to these officials, it took about 8 months for the flooding to subside. During that time, aircraft performing training and testing missions could not use the emergency runway (see fig. 4). Although intense rain events such as the one mentioned above occur periodically in the desert Southwest, DOD recognizes in its Roadmap that increases in heavy downpours are projected to increase as a result of climate change.

Source: U.S. Air Force.

Note: A dry lakebed in the Southwestern United States serves as an emergency runway for DOD training and testing missions (left); an official measures the water depth after a flash flooding event that inundated the same dry lake bed (right).

Figure 4. Dry lakebed runway in the Southwestern United States under normal and flood conditions.

Rising Sea Levels and Associated Storm Surge

At 7 out of 15 locations we visited or contacted, officials stated that they had observed rising sea levels and associated storm surge and associated potential impacts, or mission vulnerabilities, as described by DOD in the Roadmap. Officials on a Navy installation told us that sea level rise and resulting storm surge are the two largest threats to their waterfront infrastructure. For example, they explained that they were planning to lengthen a *Los Angeles*–class submarine to convert it to a training platform and that this will entail cutting the submarine in half. During this process, the submarine will sit in a dry dock with its interior open. Officials explained that they were concerned about possible storms and associated storm surge, noting that if salt water was allowed to flood the submarine's systems, it could result in severe damage. Such damage would delay completion of the submarine's lengthening by 3 to 4 months. Officials from another Navy shipyard we visited stated that flooding of a submarine in dry dock could result in catastrophic damage inside the submarine and additional, severe damage to equipment on the floor of the dry dock (see fig. 5). Thus, flooding of the submarine could result in a delay in the use of the submarine for training and fiscal exposure for DOD. Because of the perceived risk of storm surge and the possibility of the submarine's systems being exposed to flood waters, shipyard officials are considering whether to raise the current floodwall to protect the submarine.[26] According to a Navy official, raising the height of the wall is a risk-reduction strategy to address the frequency of extreme weather events happening at the shipyard.

Source: U.S. Navy.
Note: A *Los Angeles*–class submarine sits in a dry dock while it undergoes repairs.

Figure 5. Submarine in dry dock.

Changes in Ocean Temperatures, Circulation, Salinity, and Acidity

At 3 out of 15 locations we visited or contacted, officials either stated or noted that they had observed changes in ocean temperature, circulation, salinity, and acidity at their installations and that these changes may result in impacts, or mission vulnerabilities, as described by DOD in the Roadmap. For example, an official at a Navy installation explained that changes in ocean temperature may have caused a protected turtle species to nest on a section of beach where it previously had not, making this section unavailable for potential training activities during certain parts of the year. Also, the encroachment plan of a Marine Corps installation we visited states that increases in ocean temperature could lead to degradation of coral reefs in the waters offshore of the installation. As discussed above, in the Roadmap DOD associates changes in ocean temperature and acidity with potential climate change impacts such as coral-reef losses that may undermine the reef's ability to mitigate the effects of storm surge on the installation and may lead to associated mission vulnerabilities.

Installation Officials Provided Examples of Potential Climate Change Impacts That May Result in Increased Fiscal Exposure for DOD

More frequent and more severe extreme weather events and associated impacts may result in increased fiscal exposure for DOD. During our site visits, installation officials identified costs—when known—that are associated with potential impacts to infrastructure identified by DOD in its Roadmap. These costs are associated with both relatively common types of infrastructure and unique facilities. According to our discussions with officials, there is a cost to rehabilitate common types of infrastructure, such as runways and target equipment. For example, officials explained that when coastal erosion shortened a radar site's runway, DOD personnel had to access the site by helicopter, instead of the planes that personnel could have used with a full-length runway. This resulted in increased costs for DOD because reaching the site by helicopter is more expensive, according to officials.

Source: U.S. Army.

Note: Thawing permafrost results in muddy roads that reduce access to training areas (left); the same road after it was regraveled (right).

Figure 6. Road Leading to Training Area in Alaska.

In another example, when an increase in the amount of freezing rain resulted in an increased requirement for maintenance of target equipment, officials reported an added expense of target repair. Further, there are costs associated with roads and open fields impacted by thawing permafrost. For instance, officials estimate the cost to repair two gravel roads (see fig. 6) and fill holes in one drop zone to be more than $500,000. According to officials, if these types of impacts increase in frequency or severity due to climate change, DOD's routine maintenance costs are likely to increase.

Source: U.S. Air Force.

Note: Thawing permafrost, decreasing sea ice, and rising sea levels on the Alaskan coast have led to an increase in coastal erosion near a DOD early warning radar site.

Figure 7. Coastal Erosion near DOD Early Warning Site.

Officials also provided examples of costs associated with climate change impacts at unique DOD facilities. For example, officials from a Navy shipyard we visited stated that the catastrophic damage that could result from the flooding of a submarine in dry dock could cause substantial repair costs. Another example is associated with an increase in coastal erosion due to the combination of thawing permafrost, decreasing sea ice, and rising sea levels in Alaska that has damaged several types of infrastructure located near coastal radar early warning and communication facilities (see fig. 7). For instance, officials stated that due to the operational risk posed by further degradation of one facility's seawall and runway, officials are planning a project to harden the seawall and protect the runway, which they estimate will cost approximately $25 million. The erosion has also damaged landfills located near these facilities. As a consequence, landfill content—such as trash—was falling into coastal waters. These officials estimate that the cost of completing environmental restoration of six coastal sites has been $32 million thus far and more funding may be required to relocate and restore other sites.

Further, as stated in the Roadmap, extreme precipitation events may lead to potential vulnerabilities such as increased maintenance costs for roads, utilities, and runways and increased flood-control measures. For example, in August 2013, an installation located in the desert Southwest was hit with an extreme rain event in which—according to installation officials—approximately 1 year's worth of rain fell in 80 minutes. According to Army officials and documents, the flooding caused by the storm damaged more than 160 facilities—including a barracks building—8 roads, 1 bridge, and 11,000 linear feet of fencing and resulted in an estimated $64 million in damage (see fig. 8).

Source: U.S. Army Corps of Engineers Los Angeles District video footage.

Figure 8. Video Still Showing Damage to Infrastructure at Installation located in Southwestern United States.

The storm also damaged an adjacent training area used to prepare troops for deployment (see fig. 9).

Officials at the same installation told us that, in 2013, they experienced three power outages in a span of 45 days. They explained that extreme rain events caused flash floods that knocked out utility poles that transmit electricity to the installation. According to installation officials, all three of the summer 2013 power outages lasted at least 24 hours and had a substantial impact on the installation. For example, although training maneuvers could continue without electricity, the power outages limited the effectiveness of instrumentation used to track the training and provide information used for after-action feedback. Officials stated that the installation's vulnerability to power outages places community and life support systems at risk, ultimately impacting the installation's ability to meet mission training requirements. Further, officials stated that depending on when they occur in the training schedule, these types of events could impact readiness. To address this vulnerability, the installation requested more than $11.5 million to purchase 31 backup generators.

Source: GAO

Note: Guard towers at an Army training area in the Southwestern United States (left); The same type of guard tower, toppled and severely damaged by flash flooding from an extreme precipitation event at this training area (right).

Figure 9. Army Training Area in Southwestern United States.

DOD HAS BEGUN CONDUCTING VULNERABILITY ASSESSMENTS OF CLIMATE CHANGE IMPACTS ON ITS INSTALLATIONS, BUT LACK OF PLANNING MAY HAMPER COMPLETION OF ITS EFFORTS

DOD has identified potential vulnerabilities of its infrastructure to climate change impacts and has undertaken initial efforts to assess the vulnerabilities of individual installations and sites, but it lacks a plan to guide its vulnerability-assessment efforts and is not fully using interim milestones to track its progress toward completion of its goal. According to our previous work on project-management practices,[27] a project plan should guide the execution of a project, and generally accepted project-management practices emphasize the importance of identifying interim milestones.[28] As a result of not fully utilizing a project plan or interim milestones, DOD's ability to complete its vulnerability-assessment efforts in an efficient manner may be hampered, and decision makers may lack the necessary information required to assign adequate and timely resources to the most-vulnerable installations and sites.

DOD's Initial Vulnerability-Assessment Efforts Consist of Collecting Installation and Site Data and Developing Sea Level Rise Scenarios

DOD has identified examples of potential vulnerabilities of its infrastructure to climate change in selected documents, and it has started an effort to assess its infrastructure by collecting data on individual installations and sites through a survey and developing regional sea level rise scenarios.[29] DOD had previously identified examples of potential agency-wide vulnerabilities in a number of documents, including the 2010 and 2014 Quadrennial Defense Reviews, Roadmap, and an Army 2013 High-level Climate Change Vulnerability Assessment. However, according to the 2014 Quadrennial Defense Review and DOD officials, the department's goal is to complete a comprehensive assessment of all installations to assess the potential impacts of climate change on DOD's missions and to develop and implement plans to adapt as required. To meet the goal of completing a comprehensive assessment of all DOD installations, the Office of the Secretary of Defense and the military departments are, according to DOD officials, conducting screening-level assessments of 7,591 installations and sites,[30] both inside and outside of the United States.[31] According to officials from the Office of the Secretary of Defense and the military services, the goal of these assessments is to allow the services to prioritize subsequent vulnerability-assessment and adaptation actions. For example, DOD officials have suggested that— based on the vulnerabilities identified in the screening-level assessment—the military services could decide among three alternatives for individual installations and sites: (1) take no adaptation action because no significant vulnerabilities were identified, (2) conduct a more detailed assessment because moderate vulnerabilities were identified, or (3) undertake immediate adaptation actions because significant vulnerabilities were identified. Thus, DOD officials explained that the results from screening-level assessments conducted at one set of installations and sites could be used to inform adaptation decisions at these sites while other more detailed assessments are conducted at other installations and sites. In addition, officials from the Office of the Secretary of Defense stated that the screening level assessments may identify the need for revised standards both for the construction and maintenance of its infrastructure and for installation planning processes, such as master planning and natural-resources management planning. We discuss these planning processes later in this report.

DOD is conducting the screening-level assessments through two main efforts: it is collecting data on installations' and sites' historical and potential future vulnerabilities, and developing regional sea level rise scenarios, which the services intend to use following the collection of these data. In order to carry out these activities, DOD's Climate Change Adaptation Working Group formed two additional working groups. To collect data, it has formed the Assessment Guidance Sub-Working Group (Assessment Group). The Assessment Group has completed a pilot survey, through which it developed lessons learned, and used those—in part—to develop and distribute a survey that asks installation and site personnel to provide information on historical vulnerabilities to extreme weather or potential climate change impacts. For example, officials from the Office of the Secretary of Defense stated that one lesson learned is to include questions about potential climate change impacts on cultural heritage sites located on installations. The survey consists of 38 questions that, for example, ask for data on extreme temperatures the installation or site has experienced or whether flooding has impacted its infrastructure. In addition, according to DOD officials, the Climate Change Adaptation Working Group determined that it needs to develop regional sea level rise scenarios because current, global sea level rise scenarios require regional adjustments for the purposes of assessing installations' and sites' vulnerability to potential sea level rise. To develop these regional adjustments, DOD has partnered with other federal agencies, such as the National Oceanic and Atmospheric Administration, in a multiagency Coastal Assessment Regional Scenario Working Group (Scenario Group), which intends to provide these scenarios to DOD and the military services when they are complete. In addition, according to DOD officials, the services could also use these scenarios in more detailed, subsequent vulnerability assessments.

According to DOD officials, the department's intention is to conduct screening-level assessments in two phases: coastal and noncoastal. DOD has, according to officials, prioritized its initial assessment efforts on 704 coastal installations and sites. According to officials from the Office of the Secretary of Defense, the decision to prioritize coastal installations and sites was based on several considerations, including—among other factors—the high number of DOD installations and sites within 2 kilometers of a coastally or tidally influenced area,[32] recent updates to the Federal Emergency Management Agency's flood hazard maps, and the benefit that coastal assessments provide to all the military departments, given that they operate in coastal environments. DOD officials explained that after the services have used the Assessment Group's survey to collect data from the 704 coastal installations and sites, they intend to collect data from the remaining 6,887 non-coastal installations and sites for a total of 7,591 installations and sites.

In our previous work on climate change adaptation, [33] we have reported that adaptation is meant to reduce vulnerabilities to emerging or future impacts that may become more serious over time. In this regard, adaptation is a form of risk management. DOD has also emphasized the importance of risk management in its guidance for a variety of efforts, including the protection of critical infrastructure.[34] According to DOD officials and documents, one factor in the evaluation of the vulnerability of critical infrastructure is the impact of the types of extreme weather events that may become more frequent or severe due to climate change. In our discussions with DOD officials, they suggested a number of potential considerations on which to base the prioritization of future vulnerability assessments. Among these suggestions are to focus on installations and sites in the Defense Critical Infrastructure Protection Program; installations and sites located in a geographic region in which infrastructure has experienced—or may experience—relatively rapid climate change, such as the Arctic and Alaska; or installations and sites in regions of strategic priority, such as the Pacific.

DOD Has Started Initial Assessments but Lacks a Project Plan and Milestones for Completion and Risks Not Finishing the Assessments in a Timely or Complete Fashion

Although DOD has started initial screening-level assessments of its coastal installations and sites, it lacks a project plan to guide its screening-level assessment efforts and is not fully using interim milestones to track its progress toward completion of its goal. Specifically, we found that the Assessment Group has developed some interim milestones for the collection of data from the coastal installations and sites, and the Scenario Group has developed some interim milestones for completion of the regional sea level rise scenarios. However, in our discussions with officials from the Office of the Secretary of Defense and military services, neither the office nor the services were able to provide us with a project plan or a full set of milestones to guide the completion of DOD's screening-level assessment efforts in regard to both the collection and use of installation and site data and the use of regional sea level rise scenarios. For example:

- Officials from the Assessment Group were not able to provide us a plan or a full set of milestones for how they would complete the screening-level vulnerability assessment of noncoastal installations and sites. According to these officials and documentation we reviewed, the Assessment Group has established some milestones. For example, the Assessment Group has milestones for when it intends to send its survey to the 704 coastal installations and sites, and to collect these survey data. However, the Assessment Group does not have a plan to accomplish the substantial amount of work involved in collecting these data and sending surveys to a total of 7,591 installations and sites, nor does it have a full set of milestones for finishing the collection of data from the noncoastal installations and sites.
- The services were also not able to provide us with plans or milestones for how they will use the substantial amount of data collected by the Assessment Group to assess their installations' and sites' vulnerability to potential climate change impacts. Based on our review of the survey sent to coastal installations and sites and the total number of installations and sites DOD intends to assess, the services may eventually collect more than 580,000 pieces of data. Service officials generally agree that they could use these data to correlate the historical frequency of a certain impact—such as flooding—to a certain level of vulnerability. However, they have not established a methodology to conduct their suggested analysis and do not have milestones to guide any such efforts through completion.
- The military services were not able to provide us with plans for how they would use the regional sea level rise scenarios being developed by the Scenario Group. Service officials generally agree that they intend to carry out some sort of comparison between data gathered by the Assessment Group and these scenarios. However, they have not

developed a process to do so, and do not have milestones for when these comparisons will be complete.

According to our previous work that discusses project-management practices of high-performing organizations,[35] the Project Management Institute[36] has found that a project plan is essential for the success of an agency's key efforts and helps the agency attain its objectives.[37] Further, we have found in our prior work that developing and using specific interim milestones to guide and gauge progress toward achieving an agency's desired results is a leading practice for effective strategic planning and management and can help organizations more efficiently and more effectively accomplish their planned efforts.[38]

According to officials from the Office of the Secretary of Defense, DOD has several documents that reference the goals of this screening effort, including the Roadmap.[39] However these documents do not fully contain key elements of a project plan, such as project deliverables, major milestones and target dates, or required personnel. DOD officials we spoke with overseeing these efforts acknowledged that DOD does not have a project plan or full set of milestones, but believed that a written plan was not required to complete the screening-level assessments. However, according to the Project Management Institute, developing and executing a project plan are among the processes required to ensure that projects are properly coordinated. Further, a project plan is used to document project planning decisions regarding alternatives chosen by the organization. According to our discussions with DOD officials, there is currently no agreement among the organizations involved in the screening-level vulnerability-assessment process on how the remaining noncoastal assessments should be conducted. Specifically, several of these organizations are considering different alternative courses of action. For example, officials from the Office of the Secretary of Defense suggested that vulnerability assessments should be conducted concurrently at all remaining noncoastal installations and sites; officials from the Army suggested that installations and sites with military training ranges could be prioritized before other noncoastal installations and sites; and officials from the Air Force suggested that noncoastal installations and sites that contain defense critical assets should be prioritized.

Considering the substantial amount of data the services intend to collect and the time spent collecting them, the lack of a plan and milestones could prevent DOD from completing its assessments in a timely and complete manner. The Assessment Group is already several months behind its initial goal of finishing its collection of data from coastal installations and sites. The Assessment Group initially intended to finish its collection of these survey data by March 2014. However, following a delay in the distribution of the survey, the Assessment Group now intends to finish collection by June 2014. In our previous work on federal agency collaboration, we have found that in the absence of formal collaboration mechanisms—such as a project plan—informal relationships used to accomplish work can end when key personnel become unavailable.[40] According to DOD officials, the delay in distributing this survey was partly a result of similar coordination challenges related to the absence of formal collaboration mechanisms. Without timely and complete information, DOD may not be able to meet its larger goal of using assessment data to identify the need for revised construction and maintenance standards, and to inform installation master planning and natural-resources planning. Without updated planning, there could be a delay in potential adaptation actions that may be required to ensure operational readiness, as DOD recognized in its 2014 Quadrennial Defense Review.

DOD Requires Selected Installation Planning Efforts for Existing and Future Infrastructure to Account for Climate Change Impacts, but Planners Lack Key Information

Some installation planning officials have taken actions to integrate changing climatic conditions in their installation planning documents as required, but key information that would help ensure that efficient and consistent actions are taken across DOD installations has not been provided by the department. At 13 out of 15 selected locations we visited or contacted, officials stated that the relevant Unified Facilities Criteria[41] or DOD instruction 4715.03[42] did not contain key information—such as a definition of "climate change"—and they were unsure how to efficiently meet DOD's direction to address climate change in Master Plans or Integrated Natural Resources Management Plans. According to the Standards for Internal Control in the Federal Government,[43] information should be communicated in a form and within a time frame that enables personnel to carry out their responsibilities efficiently. Without additional information—such as definitions and information about projected impacts of climate change—installation planners are unlikely to consistently implement the requirements laid out in DOD guidance and consistently account for climate change impacts in their installation plans.

DOD Has Issued Guidance Requiring That Climate Change Be Considered in Specific Planning Documents

Concurrent with the screening-level vulnerability assessments discussed above, DOD has issued guidance requiring that climate change be considered in specific planning documents. According to DOD documents and officials from the Office of the Secretary of Defense, DOD's goal is to integrate consideration of climate change into existing infrastructure planning processes and documents. For example, the Roadmap states that it is DOD's goal to integrate climate change considerations into existing processes. In addition, officials from the Office of the Secretary of Defense and the military departments stated that their goal is to address potential climate change impacts and vulnerabilities through existing infrastructure planning processes so that the effects of climate change are considered in the same way that other impacts and vulnerabilities— such as for force protection—are currently considered. Executive Order 13653 from November 2013 requires several agency heads, including the Secretary of Defense, to complete an inventory and assessment of proposed and completed changes to its land- and water-related policies, programs, and regulations necessary to make the nation's watersheds, natural resources, and ecosystems more resilient in the face of a changing climate. According to officials from the Office of the Secretary of Defense, this process is currently underway within DOD, and the department has identified almost 60 planning and guidance documents for more-detailed review and potential revision.

DOD has also integrated climate change considerations into several guidance documents that provide direction to base-level planning efforts. These include DOD's Unified Facilities Criteria for Installation Master Planning[44] and DOD Instruction 4715.03.[45] The purpose of the relevant chapter of the Unified Facilities Criteria for Installation Master Planning is to, among other things, prescribe the DOD minimum requirements for master-planning processes and products. According to the document, DOD planners use the tool of a Master Plan and its components to provide ongoing master planning of installations in support of the mission, and design and programming professionals should refer to the Master Plan as they prepare site-specific design proposals for the installation. The Unified Facilities

Criteria further states that where changing external conditions impact planning decisions, master planners will seek to understand, monitor, and adapt to these changes. Such conditions include, but are not limited to, changes in climatic conditions such as temperature, rainfall patterns, storm frequency and intensity, and water levels.

In addition, DOD issued DOD Instruction 4715.03 with the purpose to—among other things—develop new policy and update policy for the integrated management of natural resources (including biological and earth resources) on property and lands managed or controlled by DOD. The instruction states that all DOD components shall, in a regionally consistent manner, and to the extent practicable and using the best science available, utilize existing tools to assess the potential impacts of climate change on natural resources on DOD installations, and, when not in conflict with mission objectives, take steps to implement adaptive management to ensure the long-term sustainability of those resources. The military departments have also started to integrate consideration of climate change into existing infrastructure planning processes. For example, according to Air Force headquarters officials, the Air Force recently completed a database that provides planners with information on certain types of environmental conditions so that planners can better integrate preparation for such events into their planning. The Air Force has also begun incorporating climate change considerations into its Installation Complex Encroachment Management Action Plans.[46] The plans involve formal assessments of 13 specific areas—one of which includes climate change—and contain prioritized action plans for current and potential issues. The Army has completed two pilot projects that it is using to help develop a planning framework for incorporating climate change considerations into existing plans and planning processes. According to an Army official, this Army Climate Change Adaptation Planning Framework will look at four types of installation plans, including Master Plans and Integrated Natural Resources Management Plans, to see where and how climate change considerations could be incorporated. It is unclear when the framework will be complete.

Installation Planners' Actions to Integrate Climate Change into Existing Planning Processes Have Varied across DOD Installations

At the 15 selected locations we visited or contacted, we found the actions planners at individual locations took to meet the requirements for Master Plans and Integrated Natural Resources Management Plans varied from location to location. For example, officials at one installation told us they directed the contractor responsible for the Master Plan to include a literature review related to climate change as part of the final Master Plan. Officials at another installation told us that because they do not consider climate change to be impacting their installation, they were not required to take further planning actions. Officials at a third installation told us they provided their master planning contractor with a study related to a single type of climate change impact, but did not have information on other types of impacts to provide to the contractor and did not know how the contractor would use the information provided by these officials. The actions planners took to meet the requirements for Integrated Natural Resources Management Plans also varied. For example, officials at one installation told us they met the requirement of DOD Instruction 4715.03 by including—in their recently updated Integrated Natural Resources Management Plan—a section discussing climate change. Officials at another installation stated that they were currently working on updating their plan, but did not anticipate including any special projects to address climate change because they are unsure what—if anything—they are expected to do. The officials told us that, instead, they will try to put already-planned projects in the context of climate change to meet the requirements.

DOD Guidance Does Not Clearly Identify Key Terms or Provide Direction Needed to Facilitate More Efficient Implementation of Guidance

Officials from 13 out of 15 selected locations we visited or contacted generally indicated that they were not sure how to efficiently implement the requirements of the Unified Facilities Criteria and DOD Instruction 4715.03 because key information is missing. For example: (1) some key terms are not defined, such as "climatic conditions;" (2) installation officials stated that updated guidance on constructing or renovating infrastructure beyond current building codes and design standards to adapt to climate change has not been provided; and (3) although DOD has identified sources of information for the potential impacts of climate change on installations (i.e., projections), certain sources contain multiple sets of information, and installation officials are not sure which information to use in their planning.

Military-service headquarters officials told us that the existing language on climate change is broadly written and acknowledged that it may pose challenges for installation planners to efficiently translate the guidance into actionable activities. For example, planners and engineers at one installation stated that they are not experts on climate change and do not feel they have the necessary expertise to define key terms such as "climatic conditions." Installation officials generally told us the documents do not define key terms, such as "climatic conditions" or "climate change," and that without the definitions they cannot be sure they are fully meeting the requirements.

Some officials also told us that current building codes and design standards in the Unified Facilities Criteria do not account for potential climate change impacts. The officials said that in order to make some changes, they would need revisions to the Unified Facilities Criteria standards that account for potential climate change impacts. They explained that the military departments build infrastructure, as required, to meet these standards. According to DOD officials, these standards require that infrastructure designs take into account historical weather phenomena and associated impacts, but the standards do not account for potential changes in these phenomena or the impacts that could result from the changes. In one example, officials told us that if the standards require that a building's roof be built to withstand 100-mph winds, simply telling an engineer that storms will "increase in intensity" is not sufficient for the engineer to design a new roof. The engineer would need a new specific wind-speed standard (120 miles per hour, for example) for which a roof should be designed.

Generally, installation officials also stated that in order to effectively consider changes in climatic conditions in their Master Plan, they would need to know—at least approximately—how the climate would change over a particular period at their installation. Some officials told us they would also need information on which species of plants and animals would be more sensitive to climate change impacts, so they could adjust their Integrated Natural Resources Management Plan accordingly. Some installation officials stated they do not feel they have the necessary expertise to determine whether climate change phenomena or impacts may be effecting their installations, or could be in the future.

Generally, officials expressed concerns about how to find or use climate change impact information. For instance, the Unified Facilities Criteria lists the National Climate Assessment and the U.S. Global Change Research Program as potential sources of information about how climate change may impact an individual installation.[47] During our discussions with installation officials, some told us they are not sure how to use the available information sources cited in the Unified Facilities Criteria in their installation-level planning efforts. One official we met with told us he planned to use one of these sources for his installation's Master Plan. However, when he learned that the website of the source did not have the information he assumed it would, he stated he was not sure what information he would use. Further, although DOD has identified sources of information for the potential impacts of climate change on installations, certain sources contain multiple sets of information, and installation officials were not sure which sets to use in their planning. For example, the Third National Climate Assessment contains regional projections for certain climate change phenomena based on two different scenarios. The scenarios differ in their assumptions about various factors that drive the rate and severity of projected climate change.[48] Officials at some installations stated that they did not know which set of projections to use in their planning efforts and would need further direction or guidance before taking action.

Some installation officials told us that because they did not have installation-level climate data from their military department or another DOD source—that may allow them to understand the potential impact of climate change on their infrastructure—they instead relied on a variety of sources. In some cases, officials relied on climate change information from non-DOD government sources, universities, contractors, and nongovernmental organizations. For example, an official at one installation explained that in the course of updating the Master Plan for her coastal installation, she provided her installation's master-planning contractor with a study on how sea level rise could impact a second coastal installation that is located hundreds of miles away from her installation. The official stated that the study—conducted by a local civic organization in the city of the second installation—was the only information she had on sea level rise. For that reason, she provided the information on the second installation to her installation's master-planning contractor for the contractor's use in addressing the Unified Facilities Criteria guidance to consider "changes in climatic conditions" for the first installation. However, the study she provided to the contractor may not have been relevant for the first installation because of the distance between the two installations and potentially other differences in the local environment around the installations. According to officials from the Office of the Secretary of Defense, local conditions specific to individual installations can have a substantial effect on how sea level rise could impact an installation. For example, in certain parts of the country, coastal land is subsiding, while in others, coastal land is rising. As a result, information on sea level rise for locations hundreds of miles away from each other may not be accurate for planning at both installations.

DOD is making efforts to compile and provide additional information to installation planners. These efforts may be positive steps, but it is too early to assess whether they will address the challenges identified by the installation officials with whom we met. As discussed previously, DOD is part of a working group in which several federal agencies are collaborating to develop regional scenarios of sea level rise, but scenarios from this working group are not yet available. According to working-group officials, the working group may have preliminary scenarios by May 2014. In addition, in November 2013, DOD issued DOD Manual 4715.03,[49] which provides guidance for incorporating climate change considerations into Integrated Natural Resources Management Plans as required by the related DOD Instruction. The new DOD manual is a positive step toward providing more direction and information that installation planners told us they need, but, at the time of our review, it was too early to assess whether the manual provides all needed information. For example, the manual does provide a definition of some key terms, but—similar to the Unified Facilities Criteria—the manual does not provide specific instructions to planners on which set of information related to future climate change impacts and associated vulnerabilities for installations they should use. According to Army officials, the Army is currently working on a pilot project at three installations to help planners assess impacts of climate change on natural resources. Army headquarters officials stated that the project will be completed by the end of fiscal year 2015 and will provide planners with additional guidance to help them integrate climate change into Integrated Natural Resources Management Plans. DOD also has research efforts underway to help provide additional information about climate change impacts to installation planners. For example, DOD is working through its Strategic Environmental Research and Development Program to conduct climate change–related studies at installations across the military departments and believes the results from the studies will help installation planners incorporate climate change consideration into their plans. In addition, researchers at the Cold Regions Research and Engineering Laboratory of the U.S. Army Corps of Engineers' Engineer Research and Development Center are studying the impact that climate change has on Alaskan installations, including on master planning and natural resource planning. As part of their work, research is being conducted at the Permafrost Tunnel Research Facility that officials stated will provide DOD and installation planners with information on how thawing permafrost can impact installations and how to adapt to any potential impacts (see figs. 10 and 11).

Source: GAO.

Note: Officials stated that research being conducted at the Permafrost Tunnel Research Facility will provide DOD and installation planners with information on how thawing permafrost can impact installations and how to adapt to potential impacts. The tunnel is approximately 15 meters below the surface and is 110 meters long.

Figure 10. Permafrost Tunnel Research Facility.

Source: GAO.

Note: Permafrost contains areas of ice that, if melted, may cause the ground above to sink. Researchers are working to help installation planners understand the impacts of thawing permafrost.

Figure 11. Ice Embedded in Soil.

According to the Standards for Internal Control in the Federal Government, information should be communicated in a form and within a time frame that enables personnel to carry out their responsibilities efficiently. Without additional information, such as the definition of "climate change" recently provided in DOD Manual 4715.03, information about projected impacts of climate change and associated vulnerabilities for installations, and updated building codes and design standards that account for potential climate change impacts, installation planners may not efficiently implement the requirements in the Unified Facilities Criteria and DOD Instruction 4715.03 and will be unlikely to consistently account for climate change impacts in their Master Plans and Integrated Natural Resources Management Plans. Further, as discussed above, installation planners may not use the most-accurate information, which could lead to inefficient adaptation planning.

DOD Processes for Approving and Funding Infrastructure Projects Do Not Explicitly Account for Climate Change Adaptation

According to DOD officials and our review of DOD documentation describing its processes for approving and funding infrastructure projects, these processes do not fully incorporate adaptation into the department's infrastructure-investment efforts. Installation officials explained that they generally have not proposed projects to address potential climate change impacts or vulnerabilities because they believe that adaptation projects will not compete well in the military services' processes for approving and funding potential projects.

This is despite a DOD strategic goal that—in its investment decisions—the department should consider sustainability, to include climate change adaptation. As a result, the military services run several risks, such as not meeting DOD's strategic goal, not effectively mitigating their installation infrastructure's potential climate change vulnerabilities, and facing more-costly adaptation actions in the future.

Installation Officials Generally Had Not Proposed Many Climate Change Adaptation Projects for Approval

According to installation and military-service headquarters officials, climate change adaptation projects for infrastructure are not widespread at the installation level. At 14 of 15 of the locations we visited or contacted, officials explained that they had not proposed or implemented a climate change adaptation project—including at 15 locations where officials had observed climate change impacts.

During our site visits, we did learn of a small number of infrastructure projects that have been started that incorporate climate change adaptation actions. In our discussions with DOD officials, we learned that a climate change adaptation project may be a stand-alone project or a component of a preexisting project. One example of a potential standalone project—discussed above—is raising the height of a floodwall at a Navy shipyard we visited. According to shipyard officials, during a recent storm, water from the storm surge flowed over the existing floodwall. Flooding such as this may pose a substantial threat to vessels and equipment in a dry dock, and a Navy official stated that the Navy is considering this floodwall project as a risk-reduction strategy to address the frequency of extreme weather events happening at the shipyard.[50]

In other cases, an adaptation action may exist as a component of a preexisting project. For instance, Air Force headquarters officials explained that at a coastal installation, the Air Force raised a building that includes a "sacrificial" first floor. As part of the existing plan to construct this building, the Air Force included this type of floor as a way to better protect critical assets—such as computer servers—from potential flooding. Specifically, critical assets could be located on the building's higher floors, while the sacrificial floor provides a less critical function, such as serving as a parking structure. If threatened by potential flooding, the parked vehicles could be moved, and if the sacrificial floor is flooded, the water would not compromise the mission capability of the building.

As discussed previously, installation planners were unsure how to efficiently meet DOD's direction to address climate change in certain types of plans, including Master Plans. Thus, they may not feel that they have the appropriate information to propose such actions. However, we have previously reported that although decision makers have imperfect information about climate change, this should not necessarily stop them from considering adaptation actions, given that adaptation actions taken now may be less expensive than actions taken later.[51]

Project Approval and Funding Processes May Limit Adaptation Projects' Ability to Compete for Funding

Installation officials explained that they generally have not proposed projects to address potential climate change impacts or vulnerabilities because they believe that adaptation projects will not compete well in the military services' processes for approving and funding potential projects. This is despite a DOD strategic goal that—in its facility-investment decisions—the department should consider sustainability, to include climate change adaptation. For example, according to its fiscal year 2012 Strategic Sustainability Performance Plan, investment decisions should consider all the costs and benefits of incorporating sustainability. Further, in the Roadmap—an appendix to this plan—DOD elucidates a goal that adaptation to climate change should not be a separate decision-making process, but rather an aspect of the overall management of existing processes, including infrastructure planning and maintenance.

Even if a potential adaptation project is included in an installation's master plan, the projects must still be approved by the military-service headquarters, before being approved by the Office of the Secretary of Defense, the Office of Management and Budget, and, ultimately, by Congress.[52] According to both our review of DOD documentation explaining typical processes and DOD officials with whom we discussed the processes, these processes—often called "scoring" processes— generally consist of the following elements: they assign numerical values—or "points"—to certain project characteristics; potential projects' relative scores are used to rank the projects; and senior decision makers at the military services' headquarters review the rank-order list, selecting projects based on service priorities.

For 14 of 15 of the locations we visited or contacted, installation officials explained that they generally have not proposed projects to address potential climate change impacts or vulnerabilities because they believe that neither standalone adaptation projects nor projects with a climate change adaptation component will score well in the military services' processes for approving and funding potential projects. For example, at an Air Force installation we visited, officials stated that because the Air Force processes for approving and funding potential projects do not explicitly account for adaptation to the impacts of climate change or extreme weather, it is unlikely that a climate change adaptation project will score highly enough to be approved. Similarly, during our visit to a Navy installation, officials shared their belief that a potential project or project component, justified as an adaptation to potential climate change impacts or mission vulnerabilities, was extremely unlikely to be approved or funded in the Navy's process. At an Army installation we visited, officials echoed this sentiment, stating that if the installation proposed a climate change adaptation project, they believe the proposal would almost certainly be denied in the Army's processes for approving and funding potential infrastructure projects.

Our review of the services' fiscal year 2013 processes for reviewing proposed military construction projects indicates that they do not explicitly account for projects intended to adapt to the potential impacts of extreme weather events or climate change. Furthermore, the factors considered in the processes—which in some cases determine how points are assigned for individual project characteristics—do not explicitly address consideration of adaptation to such impacts. Examples include the process used to compare potential military construction projects across the Air Force. This process is based on six characteristics including "Facility Condition," "Mission Dependency," and "Local Mission Impact." None of these characteristics—or the instructions associated with them— explicitly address adaptation. Similarly, a Navy process prioritizes potential military construction projects based on nine characteristics that—like the Air Force—do not explicitly include adaptation. In addition, the prioritization factors used in the Marine Corps process for approving and funding potential military construction projects include avoiding the future degradation of facilities or capabilities—but do not explicitly reference adaptation. Our review of the Army's process for approval and funding of military construction projects and the associated guidance indicates that, similarly, they do not explicitly include adaptation. Further, according to a recent draft Army report on incorporating adaptation into installation planning, projects that would improve resiliency to climate change impacts currently have no weighting in the determination of budget priorities.

Certain officials from the military service headquarters noted that in their services' processes for approving and funding potential military construction projects, projects with an adaptation component are not excluded from gaining points. For example, in discussions about the instructions associated with the Air Force characteristic "Local Mission Impact," Air Force officials explained how points could be assigned to a proposed project—according to the project's potential to help assure mission capability—and that an adaptation action has the potential do so.

Also, a Navy official explained that proposed projects could have components that serve to make a piece of infrastructure more resilient to potential climate change impacts, even if an action is not explicitly characterized as "adaptation." For example, according to Navy officials at an installation we visited, DOD plans to raise a new pier several feet above the standard height for piers, at least in part due to concerns over rising sea levels. Officials explained that they have not necessarily characterized this as a climate change adaptation action because the action is in response to observed sea level rise, as opposed to projected future changes in sea level.

However, according to military-service headquarters officials, a project that can be explicitly justified by using one of the process's established characteristics will likely gain more points than a project that cannot be explicitly justified by using one of those characteristics. These officials acknowledged that their services' processes do not explicitly address adaptation. As discussed above, installation officials with whom we met also generally believed that neither stand-alone adaptation projects nor projects with a climate change adaptation component will score well in the processes because the processes do not explicitly account for adaptation to the impacts of climate change or extreme weather.

If installation officials believe that potential adaptation projects will not be approved, the number of adaptation projects they propose is likely to be fewer than it might otherwise be, and this may result in a relatively smaller number of such projects that the services have an opportunity to approve through their processes. This poses a number of potential risks. First, the services risk not meeting DOD's strategic goal to consider sustainability, including climate change adaptation, in their facility investment decisions. Second, the services run the risk of not effectively mitigating their installation infrastructure's potential climate change vulnerabilities. In our past work examining federal efforts to aid local decision makers—such as installation level planners—in incorporating potential climate change impacts in infrastructure planning, we found that competing priorities make it difficult to pursue adaption efforts when there may be more immediate needs for attention and resources.[53] Given the potential vulnerabilities we have discussed in this report—such as installations' capability to execute their missions, potential readiness issues, and DOD's fiscal exposure to potential climate impacts—installation planners' choice not to prioritize adaptation projects could have a negative effect on DOD infrastructure and the missions it supports. Third, the military services run the risk of more-costly adaptation actions in the future. According to both the Third National Climate Assessment and our previous work on climate change adaptation, preparing for the impacts of climate change now may reduce the need for more-costly steps in the future.[54] DOD typically requests—and Congress funds—hundreds of military construction projects each year, some of which have expected life spans of more than 20 years. By not fully considering the potential adaptation needs of this new infrastructure in its current facility investment decisions, DOD may be at risk of having to make more-costly adaptation upgrades in the future.

CONCLUSION

DOD's infrastructure provides facilities and training ranges that are vital to the department's ability to fulfill its mission and constitutes a significant fiscal commitment that requires ongoing investment to maintain. According to the Roadmap, changes in the Earth's climate could include rising temperatures and sea levels, changes in precipitation patterns, and increased severity and frequency of extreme weather events. Given these potential changes, DOD's stated approach is to minimize the vulnerabilities of its infrastructure to the impacts of these phenomena in order to maintain mission readiness. This may also reduce DOD's fiscal exposure to the effects of climate change. The department's initial efforts have been reasonable responses to these challenges, given some continued uncertainty in climate science and competing departmental priorities. Still, as DOD considers next steps, it would benefit by taking actions to ensure that ongoing and future infrastructure projects are designed to be consistent with DOD's sustainability goals. For instance, DOD's Climate Change Adaptation Working Group, in deciding to focus its initial screening-level assessment efforts on coastal installations and sites, recognized that such infrastructure forms a substantial commitment of the department. And, the Climate Change Adaptation Working Group has taken some positive steps, including its intention to leverage lessons learned from the DOD's initial coastal efforts into how to finish subsequent vulnerability assessments, and its use of a risk-based approach as it prioritizes these vulnerability-assessment efforts. However, without a project plan in place that includes milestones, the department may not be able to effectively implement actions to meet its goal of carrying out vulnerability assessments for all its facilities. This could slow the implementation of adaptation actions based on vulnerabilities identified in these assessments. For example, a delay in collecting these data may, in turn, delay decision makers' use of these data when considering revised building standards and planning processes. Similarly, without additional information from the military departments that clarifies how to incorporate climate change adaptation actions into installation planning documents, installation planners may not efficiently meet DOD requirements, which could impact DOD, because the department may not have the information it needs to make informed tradeoffs when it makes infrastructure decisions. Finally, without military-service processes that ensure that climate change adaptation projects are proposed by facility planners and will be competitive for funding—when appropriate—with other projects, DOD risks missing an opportunity to enhance the resilience of its infrastructure to the impacts of climate change at the appropriate time.

RECOMMENDATIONS FOR EXECUTIVE ACTION

In order to provide DOD decision makers with comprehensive information on the vulnerabilities of its facilities to the potential impacts of climate change on its missions, we recommend that the Secretary of Defense direct the Senior Sustainability Officer and Climate Change Adaptation Working Group to—through the Assessment Group—develop a project plan and milestones for completing DOD's screening-level vulnerability assessment and direct the services to—in coordination with the Assessment Group—develop plans and milestones that describe how they intend to use the data collected through the assessment in support of climate change adaptation planning.

In order to facilitate the efforts of installation planners to efficiently implement the requirements of the Unified Facilities Criteria and DOD Instruction 4715.03, we recommend that the Secretary of Defense—in conjunction with the Secretaries of the military departments—provide further direction and information that clarifies the planning actions that should be taken to account for climate change in installation Master Plans and Integrated Natural Resource Management Plans. At a minimum, further direction could include definitions of key terms, such as the definition of "climate change" recently included in DOD Manual 4715.03; further information about changes in applicable building codes and design standards that account for potential climate change impacts; and further information about potential projected impacts of climate change for individual installations.

In order to improve the military services' ability to make facility investment decisions in accordance with DOD's strategic direction to include climate change adaptation considerations and additionally, to demonstrate an emphasis on proposing projects with an adaption component to installation planners, we recommend that the Secretary of Defense direct the Secretaries of the military departments to clarify instructions associated with the processes used to compare potential military construction projects for approval and funding so that, at a minimum, climate change adaptation is considered as a project component that may be needed to address potential climate change impacts on infrastructure.

AGENCY COMMENTS AND OUR EVALUATION

We provided a draft of this report to the Department of Defense (DOD) for review and comment. DOD provided written comments, concurring with all three of our recommendations and noting actions it intends to take to address them. In addition, the Office of the Secretary of Defense and the military departments provided technical comments. We also provided a draft of this report for review and comment to the Executive Director of the United States Global Change Research Program. The United States Global Change Research Program did not provide official written comments but instead provided technical comments. We incorporated the technical comments from the Office of the Secretary of Defense, the military departments and the United States Global Change Research Program, as appropriate.

DOD concurred with our recommendation to develop a project plan and milestones for both completing DOD's screening-level vulnerability assessment and using data collected through the assessment in support of climate change adaptation planning. DOD stated that its Assessment Guidance Sub-Working Group will document its current plans to complete the assessment, including a phased approach to surveying all the sites and taking steps to address analysis of data collected through these surveys. We believe that documenting the Assessment Group's current plans for completing the assessment, detailing how data collected through the assessment will be used in support of climate change adaptation planning, and including milestones in this documentation will better ensure that DOD decision makers have comprehensive information on facilities' vulnerabilities to the potential impacts of climate change on DOD's missions.

DOD also concurred with our recommendation to provide further direction and information that clarifies the planning actions that should be taken to account for climate change in installation Master Plans and Integrated Natural Resource Management Plans, including providing further information about potential projected impacts of climate change for individual installations. DOD stated that it has begun drafting a DOD directive on climate change adaptation that, when complete, will provide guidance and direction and define terminology, among other things. We believe that a DOD directive on climate change adaptation would be a positive step toward providing installation planners with further direction and information. DOD also stated that it is participating in the efforts of the United States Global Change Research Program to provide the U.S. government with climate change adaptation data on the data.gov climate website.[55] We believe that the website may serve as a useful source of data about certain potential projected impacts of climate change. However, as of May 2014, the website was in a pilot phase, in which the data and resources available were limited to coastal flooding, sea level rise, and their impacts. As we discuss in the report, DOD describes five categories of climate change phenomena in its fiscal year 2012 Climate Change Adaptation Roadmap and during our site visits, installation officials identified potential climate change impacts or vulnerabilities associated with each of these phenomena. If DOD plans to use this website as a source of climate change data for installation planners, and the site provides information on all types of climate change phenomena, this will meet the intent of our recommendation.

Finally, DOD concurred with our recommendation to clarify instructions associated with the processes used to compare potential military construction projects for approval and funding so that, at a minimum, climate change adaptation is considered as a project component that may be needed to address potential climate change impacts on infrastructure. DOD stated that climate change may be one of many factors that can affect facilities and impact mission and readiness, and that the department will review processes and criteria, such as the Unified Facilities Criteria, to strengthen consideration of climate change adaptation. We believe that such a review could lead to clarification of instructions associated with these processes.

Brian J. Lepore
Director, Defense Capabilities and Management

APPENDIX I: SCOPE AND METHODOLOGY

To describe the Department of Defense's (DOD's) efforts to identify climate change phenomena and associated impacts and vulnerabilities on its infrastructure, we collected information from the Office of the Secretary of Defense and the military departments and reviewed DOD documents such as the 2010 and 2014 Quadrennial Defense Reviews; DOD's fiscal year 2012 Strategic Sustainability Performance Plan; and the fiscal year 2012 Climate Change Adaptation Roadmap (Roadmap). We also reviewed ongoing studies on potential climate change impacts at DOD locations from DOD's Strategic Environmental Research and Development Program and met with officials from the Office of the Secretary of Defense and the military department headquarters.

To gain insight into examples of climate change phenomena that may have or may impact installation infrastructure in the future and that may lead to mission vulnerabilities, we visited or contacted officials at a nongeneralizable sample of 15 DOD locations,[1] learned about or observed impacts to infrastructure, and collected key documentation such as after action reports describing impacts and mission vulnerabilities of climate change phenomena. These locations were within the 50 U.S. states, the District of Columbia, the Commonwealth of Puerto Rico, Guam, the Virgin Islands, American Samoa, and any other territory or possession of the United States. We then selected locations that represented a variety of geographic regions—regions used by both DOD and the United States Global Change Research Program's National Climate Assessment when discussing climate change[2]—as well as locations from each of the military departments. The locations we selected were locations that DOD has identified as already being impacted by climate change, or those that DOD is expecting to be impacted by climate change. To ensure this, we selected locations where either (1) research on climate change had been conducted, or was planned to be conducted, by either the Office of the Secretary of Defense's Strategic Environmental Research and Development Program or the military departments; or (2) DOD officials had identified as being impacted, or likely to be impacted, by climate change. We selected, and visited or contacted the following installations: Combat Direction Systems Activity, Naval Air Station Oceana, Dam Neck Annex, Virginia; Edwards Air Force Base, California; Fort Irwin, California; Fort Irwin National Training Center, California; Fort Wainwright, Alaska; Joint Base Pearl Harbor–Hickam, Hawaii; Joint Pacific Alaska Range Complex, Alaska; Marine Corps Base Hawaii, Hawaii; Naval Air Station Norfolk, Virginia;

Naval Air Station Oceana, Virginia; Naval Base Guam, Guam; Norfolk Naval Shipyard, Virginia; Pearl Harbor Naval Shipyard and Intermediate Maintenance Facility, Hawaii; Yukon Training Area, Alaska; and 611th Civil Engineer Squadron remote radar sites, Alaska. Results from our nongeneralizable sample cannot be used to make inferences about all DOD locations. However, the information from these locations provides valuable insights.

To determine the extent to which DOD has assessed or plans to assess the vulnerability of its facilities to the potential impacts of climate change, we reviewed DOD's 2010 and 2014 Quadrennial Defense Reviews and Roadmap to determine DOD' goals for conducting vulnerability assessments. To determine the extent to which the department has completed such assessments, we reviewed DOD documentation that discusses vulnerability assessments and interviewed officials in the Office of the Secretary of Defense and military departments to learn about DOD's assessment efforts. In addition, we observed meetings of DOD's key organizations involved in these assessment efforts, including the Adaptation Group, the Assessment Group, and the Scenario Group. Finally, we reviewed project-management practices of high-performing organizations, comparing DOD's assessment efforts to these practices in order to determine the extent to which DOD has incorporated them into its assessment efforts.

To determine the extent to which DOD is accounting for climate change impacts in installations' selected planning efforts for existing and future infrastructure, we identified the guidance for doing so as it relates to Installation Master Planning and Integrated Natural Resources Management Plans. We then compared these directions with actual Master Plans and Integrated Natural Resources Management Plans for locations we visited or contacted, and we met with officials from those locations who were responsible for the Master Plan and Integrated Natural Resources Management Plans to learn about how they incorporated the guidance from DOD into their planning processes. We also reviewed the Standards for Internal Control in the Federal Government related to information and communications,[3] and compared those standards to DOD's practices related to Installation Master Planning and Integrated Natural Resources Management Plans.

To determine the extent to which DOD accounts for climate change impacts when making infrastructure project funding decisions, we reviewed DOD's fiscal year 2012 Strategic Sustainability Performance Plan and Roadmap to determine DOD's strategic goals for integrating climate change adaptation actions into facility investment decision processes. We also interviewed DOD officials at the military-department headquarters level and at the selected installations we visited or contacted to determine the extent to which the military departments have implemented climate change adaptation efforts at the installation level. In addition, we reviewed examples of military department processes used for approval and funding of potential infrastructure projects. Finally, we compared the extent to which the processes incorporate climate change adaptation with DOD's strategic goals on integrating adaptation actions into facility investment decisions.

APPENDIX II: PREVIOUS GAO WORK ON FEDERAL EFFORTS TO ADDRESS CLIMATE CHANGE HAS RESULTED IN A NUMBER OF RECOMMENDATIONS FOR IMPROVEMENT OF THESE EFFORTS

In February 2013, we added the federal government's efforts to manage its fiscal exposure to the effects of climate change to our High Risk list, noting that the federal government's role as a property owner exposes it to significant fiscal risk. This is the first GAO report focused on DOD climate change adaptation efforts for infrastructure. However, GAO has a body of work on climate change that—among other topics—addresses the impacts of climate on infrastructure and assesses adaptation efforts across the federal sector.

In a 2009 report,[1] we noted that climate change has implications for the fiscal health of the federal government because it puts new stresses on the infrastructure that federal agencies manage. To address this challenge, we recommended a national strategic plan that defines federal priorities related to adaptation and addresses how resources will be made available to implement them. The Deputy Associate Director for Climate Change Adaptation at the Council on Environmental Quality generally agreed with the recommendations of the report, noting that leadership and coordination is necessary within the federal government to ensure an effective and appropriate adaptation response and that such coordination would help to catalyze regional, state, and local activities.[2]

In a 2013 report,[3] we stated that extreme weather events and climate change pose risks to physical infrastructure and discussed observed climate change impacts in the United States. Specifically, we noted that— according to assessments by the National Research Council and the United States Global Change Research Program—changes in the climate have been observed in the

United States and its coastal waters and are projected to grow in severity in the future, thereby increasing the vulnerability of infrastructure. We found that decision makers had not systematically incorporated climate change impacts into infrastructure planning because, among other reasons, available climate change information did not easily fit into their infrastructure planning process.

Therefore, we noted that the federal government could provide technical assistance to help decision makers translate available climate-related data into information useful for decision making and recommended that the Executive Director of the United States Global Change Research Program or other federal entity designated by the Executive Office of the President (1) work with relevant agencies to identify for decision makers the "best available" climate-related information for infrastructure planning and update this information over time, and (2) clarify sources of local assistance for incorporating climate-related information and analysis into infrastructure planning, and communicate how such assistance will be provided over time. Further, we recommended that the Secretary of Transportation and the Administrator of the Environmental Protection Agency work with relevant professional associations to incorporate climate change information into design standards. Relevant entities provided technical comments but did respond to these recommendations.

End Notes

[1] For the purposes of this report, we define infrastructure as all buildings and permanent installations necessary for the support, redeployment, and operations of (e.g., barracks, headquarters, airfields, communications facilities, stores, port installations, and maintenance stations). Infrastructure includes utility systems; training and testing ranges and areas; and transportation systems (e.g., roads, bridges). Also, to the extent that DOD officials considered built or natural infrastructure outside of a facility (e.g., utility lines or barrier islands, respectively) in their climate-change adaptation planning for the facility, we included this infrastructure in the scope of the report.

[2] According to DOD, climate change is any given change in climate over time, whether due to natural variability or as a result of human activity.

[3] GAO, *Extreme Weather Events: Limiting Federal Fiscal Exposure and Increasing the Nation's Resilience*, GAO-14-364T (Washington, D.C.: Feb. 12, 2014).

[4] The Third National Climate Assessment was released in May 2014.

[5] The U.S. Global Change Research Program coordinates and integrates the activities of 13 federal agencies that conduct research on changes in the global environment and their implications for society. The Global Change Research Program began as a presidential initiative in 1989 and was established by law in the Global Change Research Act of 1990 (Pub. L. No. 101-606, § 103 (1990)). The U.S. Global Change Research Program

participating agencies are the Departments of Agriculture, Commerce, Defense, Energy, Interior, Health and Human Services, State, and Transportation; the U.S. Agency for International Development; the Environmental Protection Agency; the National Aeronautics and Space Administration; the National Science Foundation; and the Smithsonian Institution.

[6] Limiting the Federal Government's Fiscal Exposure by Better Managing Climate Change Risks in GAO, *High Risk Series: An Update*, GAO-13-283 (Washington, D.C.: Feb. 14, 2013).

[7] Section 118 of Title 10 of the United States Code requires the Secretary of Defense to conduct a comprehensive examination of the national defense strategy, force structure, force modernization plans, infrastructure, budget plan, and other elements of the defense program and policies of the United States, every 4 years, with a view toward determining and expressing the nation's defense strategy and establishing a defense program for the next 20 years.

[8] DOD's *Roadmap* is an annex to its Fiscal Year 2012 *Strategic Sustainability Performance Plan.* DOD developed the *Roadmap* in response to Executive Order 13514 on *Federal Leadership in Environmental, Energy, and Economic Performance* that directs federal agencies to evaluate their climate change risks and vulnerabilities to manage the effects of climate change on the agency's operations and mission in both the short and long term.

[9] For the purposes of this report, the United States consists of the 50 states, the District of Columbia, the Commonwealth of Puerto Rico, Guam, the Virgin Islands, American Samoa, and any other territory or possession of the United States.

[10] Department of Defense, Unified Facilities Criteria 2-100-01, *Installation Master Planning* (May 15, 2012).

[11] Department of Defense Instruction 4715.03, *Natural Resources Conservation Program* (Mar. 18, 2011).

[12] GAO, *Standards for Internal Control in the Federal Government,* GAO/AIMD-00-21.3.1 (Washington, D.C.: November 1999).

[13] GAO, *Climate Change Adaptation: Strategic Federal Planning Could Help Government Officials Make More Informed Decisions*, GAO-10-113 (Washington, D.C.: Oct. 7, 2009).

[14] Major greenhouse gases include carbon dioxide (CO2); methane (CH4); nitrous oxide (N2O); and synthetic gases such as hydrofluorocarbons (HFC), perfluorocarbons (PFC), and sulfur hexafluoride (SF6).

[15] Other organizations include the National Research Council and the U.S. Global Change Research Program. The National Research Council is the operating arm of the National Academy of Sciences and the National Academy of Engineering. The U.S. Global Change Research Program coordinates and integrates the activities of 13 federal agencies that conduct research on changes in the global environment and their implications for society.

[16] Global climate models are global, three-dimensional computer models of the climate system that can be used to simulate climate change. The models are highly complex and they represent the effects of such factors as absorptive properties of atmospheric water vapor, greenhouse gas concentrations, clouds, annual and daily solar heating, ocean temperatures, and ice boundaries. The most-recent global climate models include global representations of the atmosphere, oceans, and land surface.

[17] Executive Order 13514, *Federal Leadership in Environmental, Energy, and Economic Performance* (Oct. 5, 2009).

[18] *Instructions for Implementing Climate Change Adaptation Planning in Accordance with Executive Order 13514 "Federal Leadership in Environmental, Energy, and Economic Performance,"* (Mar. 4, 2011).

[19] Executive Order 13653, *Preparing the United States for the Impacts of Climate Change* (Nov. 1, 2013).

[20] GAO recently reported on maritime infrastructure related to commercial activity in the U.S. Arctic. GAO, *Maritime Infrastructure: Key Issues Related to Commercial Activity in the U.S. Arctic over the Next Decade*, GAO-14-299 (Washington, D.C.: Mar. 19, 2014).

[21] We have reported that while climate change is typically described in terms of average annual changes in temperature or precipitation, it is also associated with shifts in the frequency and severity of extreme weather events to which physical infrastructure is particularly vulnerable. See GAO, *Climate Change: Future Federal Adaptation Efforts Could Better Support Local Infrastructure Decision Makers*, GAO-13-242 (Washington, D.C.: Apr. 12, 2013).

[22] Officials at several locations told us that they observed multiple types of what DOD has identified as potential climate change phenomena, potential impacts, or potential vulnerabilities.

[23] According to Air Force officials, overrun areas at the end of runways serve as emergency space to slowly stop planes that overrun the runway during landing or takeoff.

[24] According to DOD, a drop zone is a specific area upon which airborne troops, equipment, or supplies are air-dropped.

[25] According to the U.S. Army Corps of Engineers, Cold Regions Research and Engineering Laboratory, permafrost is soil or rock, or both, that has remained below 32°F for more than 2 years. Alaska is covered by different types of permafrost: (1) continuous, where more than 80 percent of the ground surface is underlain by permafrost (found in very high latitudes); (2) discontinuous, where permafrost exists in certain areas depending on the vegetation, soil type, moisture, and exposure to solar radiation (Fairbanks is in the discontinuous permafrost zone); and (3) sporadic, where less than 30 percent of the ground surface is underlain by permafrost.

[26] Officials told us that they are planning to raise the floodwall to withstand a "500-year flood." According to the U.S. Geological Survey, this type of flood has a 0.2 percent chance of occurring in any given year. The current floodwall is designed to withstand a "100-year flood," or a flood with a 1 percent chance of occurring in any given year. According to the U.S. Geological Survey, the term "100-year flood" is a statistical designation, meaning that there is a 1-in-100 chance that a flood this size will happen during any year. According to installation officials, the shipyard typically prepares for a 100-year flood. Therefore, preparing for a 500-year flood represents preparation for more-severe flooding.

[27] For example, see our previous work evaluating the use of project-management practices in the modernization of information-technology systems at the Department of Housing and Urban Development. GAO, *Information Technology: HUD Needs to Improve Key Project Management Practices for Its Modernization Efforts*, GAO-13-455 (Washington, D.C.: June 12, 2013). For the purposes of this report, a project plan is defined as a formal, approved document used to guide both project execution and project control. The primary uses of the project plan are to document planning assumptions and decisions, facilitate communication among stakeholders, and document approved scope, cost, and schedule baselines.

[28] GAO, *Coast Guard: Civil Rights Directorate's Action Plans to Improve Its Operations Could Be Strengthened by Implementing Several Aspects of Project Planning and Implementation Practices*, GAO-10-571T (Washington, D.C.: Apr. 27, 2010). For the purposes of this report, milestones are defined as significant events in the project, usually completion of a major deliverable.

[29] According to the Third National Climate Assessment, scenarios are ways to help understand what future conditions might be, with each scenario an example of what might happen under particular assumptions.

[30] DOD's Assessment Group has identified 523 installations for which vulnerability-assessments should be conducted. For the purposes of these assessments, the Assessment Group has further identified over 7,000 noncontiguous sites operated by these installations. According to an official from the Office of the Secretary of Defense, the Assessment Group is collecting data from a total of 7,591 installations and sites. The Assessment Group originally intended to collect data from almost 11,700 installations and sites. However, as a result of discussions with the GAO, DOD reviewed its list of these installations and sites. Through the review, DOD officials determined that almost 4,100 Army Corps of Engineers' civil works sites included in the original count of 11,700 should not have been included. In addition, DOD revised its list of installations and sites to include almost 130 Washington Headquarters Service sites, as well as updated numbers for National Guard, Reserves, and recruiting sites. As a result of this review, the Assessment group now intends to collect data from a total of 7,591 sites.

[31] According to an official from the Office of the Secretary of Defense, more than 90 percent of the installations and sites that DOD intends to survey are located in the United States.

[32] According to an official from the Office of the Secretary of Defense and a U.S. Army Corps of Engineers regulation, coastally or tidally influenced areas are locations with oceanic astronomical tidal influence and connected waterways with base water levels controlled by sea level.

[33] GAO-13-242.

[34] See Department of Defense Instruction 3020.45, *Defense Critical Infrastructure Program (DCIP) Management* (Apr. 21, 2008).

[35] GAO-13-455.

[36] According to its website, the Project Management Institute is a not-for-profit association that, among other things, seeks to improve organizations' success in project management.

[37] GAO-13-455.

[38] We have previously reported that without interim milestones, decision makers may lack the information they need to assess progress and estimate realistic completion dates. See, for example, GAO, *Managing For Result: Agencies Should More Fully Develop Priority Goals under the GPRA Modernization Act*, GAO-13-174 (Washington, D.C.: Aug. 19, 2013).

[39] According to officials from the Office of the Secretary of Defense, other documents that contain this goal include both the fiscal year 2010 and fiscal year 2014 Quadrennial Defense Reviews; the Assessment Group's charge document; and an internal tasking memo sent by the Office of the Secretary of Defense to the military services.

[40] GAO, *Environmental Health: Action Needed to Sustain Agencies' Collaboration on Pharmaceuticals in Drinking Water*, GAO-11-346 (Washington, D.C.: Aug. 8, 2011).

[41] Department of Defense, Unified Facilities Criteria 2-100-01, *Installation Master Planning* (May 15, 2012).

[42] Department of Defense Instruction 4715.03, *Natural Resources Conservation Program* (Mar. 18, 2011).

[43] GAO/AIMD-00-21.3.1

[44] Department of Defense, Unified Facilities Criteria 2-100-01, *Installation Master Planning* (May 15, 2012).

[45] Department of Defense Instruction 4715.03, *Natural Resources Conservation Program* (Mar. 18, 2011).

[46] According to the Air Force, encroachment is defined as any deliberate action by any governmental or nongovernmental entity or individual that does, or is likely to, inhibit, curtail, or impede current or future military activities within the installation complex or mission footprint, or both; or deliberate military activity that is, or is likely to be, incompatible with the use of a community's resources.

[47] The Unified Facilities Criteria refers to the U.S. Global Change Research Office. According to Office of the Secretary of Defense officials, installation planners should reference the U.S. Global Change Research Program. The officials did not have an explanation for the discrepancy.

[48] In the Third National Climate Assessment, projections are based on simulations of the 21st-century climate that assume changing greenhouse gas concentrations following two scenarios: A2 (high emissions) and B1 (low emissions).

[49] Department of Defense Manual 4715.03, *Integrated Natural Resources Management Plan (INRMP) Implementation Manual* (Nov. 25, 2013).

[50] As discussed above, although installation officials generally agreed that individual weather events—such as a single storm—cannot be tied directly to changes in the climate, these officials noted that DOD facilities and infrastructure are vulnerable to climate change phenomena and recognized that climate change may make these types of phenomena more frequent or severe. Also, as discussed above, we have previously reported that while it is not possible to link any individual weather event to climate change, these events provide insight into the potential climate-related vulnerabilities the United States faces (GAO-14-364T).

[51] GAO-13-242.

[52] For this report, we did not review the portions of the approval process that occur after approval by the military services.

[53] GAO-13-242.

[54] GAO-13-242.

[55] For more information, see http://www.data.gov/climate

End Notes for Appendix I

[1] We visited 13 of the 15 locations. During our site visits, we discussed the other 2 locations.

[2] We selected locations in the following regions: Southwest, Southeast and Caribbean, Hawaii and Pacific Islands, and Alaska and the Arctic. We did not select locations in: Northeast, Midwest, Great Plains, or Northwest.

[3] GAO, *Standards for Internal Control in the Federal Government,* GAO/AIMD-00-21.3.1 (Washington, D.C.: November 1999).

End Notes for Appendix II

[1] GAO, *Climate Change Adaptation: Strategic Federal Planning Could Help Government Officials Make More Informed Decisions*, GAO-10-113 (Washington, D.C.: Oct. 7, 2009).

[2] The Council on Environmental Quality was established within the Executive Office of the President by Congress as part of the National Environmental Policy Act of 1969. Pub. L. No. 91-190 (1970) (codified as amended at 42 U.S.C. § 4321, et seq.).

[3] GAO, *Climate Change: Future Federal Adaptation Efforts Could Better Support Local Infrastructure Decision Makers*, GAO-13-242 (Washington, D.C.: Apr. 12, 2013).

In: Climate Change
Editor: Ezra Stack

ISBN: 978-1-63463-383-3

Chapter 2

DEPARTMENT OF DEFENSE FISCAL YEAR 2012 CLIMATE CHANGE ADAPTATION ROADMAP*

U.S. Department of Defense

The Department of Defense (DoD) Climate Change Adaptation Roadmap (CCAR) fulfills a requirement of Executive Order 13514, *Federal Leadership in Environmental, Energy and Economic Performance.* All Federal Departments and Agencies should evaluate climate change risks and vulnerabilities to manage both the short- and long-term effect of climate change on the agency's mission and operations, and include an adaptation planning document as an appendix to its annual Strategic Sustainability Performance Plan.

1. POLICY FRAMEWORK FOR CLIMATE CHANGE ADAPTATION PLANNING

The foundation for DoD's strategic policy on climate change adaptation began with the publication of the Quadrennial Defense Review (QDR) in 2010

* This is an edited, reformatted and augmented version of material that originally appeared as Appendix 2 in the Department of Defense's *Strategic Sustainability Performance Plan*, issued in September 2012.

by the Secretary of Defense. The QDR is a principal means by which the tenets of the National Defense Strategy are translated into new policies and initiatives. The QDR sets a long-term course for DoD as the Department assesses the threats and challenges that the nation faces and re-balances DoD's strategies, capabilities, and forces to address today's conflicts and tomorrow's threats. The QDR acknowledged that climate change has national security implications and must be addressed by DoD and its partners.

The QDR recognized that climate change will affect DoD in two broad ways.

- First, climate change will shape the operating environment, roles, and missions that the Department undertakes. It may have significant geopolitical impacts around the world, contributing to greater competition for more limited and critical life-sustaining resources like food and water. While the effects of climate change alone do not cause conflict, they may act as accelerants of instability or conflict in parts of the world. Climate change may also lead to increased demands for defense support to civil authorities for humanitarian assistance or disaster response, both within the United States and overseas.
- Second, DoD will need to adjust to the impacts of climate change on its facilities, infrastructure, training and testing activities, and military capabilities. DoD's operational readiness hinges on continued access to land, air, and sea training and test space, all of which are subject to the effects of climate change.

Through its planning and adaptation actions, DoD will be better prepared to effectively respond to climate change and to ensure continued mission success, both in the near term and in the future.

> "Our mission at the Department is to secure this nation against threats to our homeland and to our people. In the 21st Century, the reality is that there are environmental threats which constitute threats to our national security. For example, the area of climate change has a dramatic impact on national security: rising sea levels, to severe droughts, to the melting of the polar caps, to more frequent and devastating natural disasters all raise demand for humanitarian assistance and disaster relief."
>
> *Secretary Leon F. Panetta, May 2, 2012.*

As climate science advances, the Department will need to regularly reevaluate climate-related risks and opportunities in order to develop policies and plans that manage climate change's impacts on the Department's operating environment, missions, and facilities. Managing the national security implications of climate change will require DoD to work collaboratively, with both traditional allies and new partners.

1.A. Vision and Goals

As articulated in the Department's Strategic Sustainability Performance Plan, DoD's sustainability vision is to maintain our ability to operate into the future without decline, either in the mission or the natural and man-made systems that support it. Including climate change and climate variability considerations in our planning processes will enhance operational and infrastructure resilience.

Four broad goals support the Department's vision, as detailed below; implementation is discussed in Section 3.

1. Define a coordinating body to address climate change.
2. Utilize a robust decision making approach based on the best available science.
3. Integrate climate change considerations into existing processes.
4. Partner with Federal agencies and allies on the challenges of climate change.

> "Our ability to advance constructive cooperation is essential to the security and prosperity of specific regions, and to facilitating global cooperation on issues ranging from violent extremism and nuclear proliferation, to climate change, and global economic instability-issues that challenge all nations, but that no one nation alone can meet." (p. 11)
>
> — 2010 National Security Strategy.

1.B. Responsible Senior Agency Official

The Department's Senior Sustainability Officer (SSO) is the Under Secretary of Defense for Acquisition, Technology, and Logistics (USD(ATL)) and is responsible for overseeing the implementation of the requirements of Executive Order 13514, including climate change adaptation efforts. The Deputy Under Secretary of Defense for Installations and Environment (DUSD(I&E)) is the Department's Senior Climate Official and reports progress to the SSO. Given the broad range of potential impacts to the Department's operational, training, and test and infrastructure capabilities, the Department will analyze how climate change adaptation measures can be incorporated into the full scope of its missions and operations.

2. AGENCY VULNERABILITY: ANALYSIS OF CLIMATE CHANGE RISKS AND OPPORTUNITIES

Climate change is expected to play a significant role in DoD's ability to fulfill its mission in the future. Climate-related effects already are being observed at DoD installations throughout the U.S. and overseas. The physical changes are projected to include rising temperature and sea level and increases in both heavy downpours and the extent of drought. These will cause effects such as more rapid coastal erosion, shifts in growing seasons, and changing water tables.

The direction, degree, and rates of the physical changes will differ by region, as will the impacts to the military's mission and operations. By taking a proactive, flexible approach to vulnerability assessment and adaptation planning that recognizes uncertainty and incorporates the best available science, the Department can keep pace with changing climate patterns and minimize their impact on operations.

The military is potentially vulnerable to climate change in many of the same ways as the rest of society, and in ways that are unique due to its operations and mission. The following table summarizes the potential high-level climate change impacts to the Department's mission and operations. More comprehensive and region/installation-specific vulnerability assessments are needed to determine what adaptive responses are the most appropriate.

Table 1. Climate Phenomena and Potential DoD Mission Vulnerabilities

Climate Change Phenomena	Potential Impacts	Potential Mission Vulnerabilities
Rising temperatures	Rising mean temperatures; seasonal temperature increases;	Increased occurrence of test/training limitations due to

Climate Change Phenomena	Potential Impacts	Potential Mission Vulnerabilities
	increased number of cumulative days with temperatures exceeding 95^{0}F; opening of Arctic waters; melting permafrost and ice sheets; lengthening ice-free seasons; human health effects; vegetation transition (species and biome shifts); changes in incidence/ distribution of vector-borne diseases; wildfire risk; soil warming; electrical grid stress; equipment performance	high heat days; reduced military vehicle access (e.g., melting permafrost); degrading infrastructure and increased maintenance costs for roads, utilities, and runways; reduced airlift capacity; reduced live-fire training; potential degradation or loss of cold weather training venues; increased energy costs for building and industrial base operations; increased operational health surveillance and risks; change in operational parameters for weapons and equipment development and testing; increase in seasonal Arctic commerce and transit
Changes in precipitation patterns	Seasonal increases and decreases in precipitation; increases in extent and duration of drought; increases in extreme precipitation events; changes in number of consecutive days of high or low precipitation; change in form of precipitation (i.e., snow-ice-rain); increased wildfire risk; altered burn regimes; impacts to air quality; stream bank erosion and gullying of vegetative cover; impacted soil function and resilience (desertification); soil loss; infrastructure damage; water supply constraints; impacted groundwater quality; increased dust; protected species stress and potential for more species placed at risk; spread of invasive species; changes in incidence/distribution of vector-borne diseases; land management impacts; competing non-military land use	Reduced land carrying capacity for vehicle maneuvers; increased maintenance costs for roads, utilities, and runways; limits on low-level rotary wing flight operations; icing on aircraft; increased regulatory constraints on training land access; reduced live-fire training; reduced water availability and greater competition for limited water resources; reduced training land access; reduced training carrying capacity; operational health surveillance and risks; increased flood control/erosion prevention measures

Table 1. (Continued)

Climate Change Phenomena	Potential Impacts	Potential Mission Vulnerabilities
Increasing storm frequency & intensity (coastal and inland) Rising sea levels & associated storm surge	Flooding; water quality issues; soil and vegetation loss; impact to soil function and carbon/nutrient cycling; wind damage Loss of coastal land; damage to physical infrastructure (roads, targets, ranges) and protected ecosystem resources; saltwater intrusion; reduced capacity of protective barrier islands and coastal wetlands	Military personnel safety; temporary or prolonged disruption of military operations or test and training activities due to intense storms and resulting storm damage; inundation of and damage to coastal infrastructure; reduced access to military water crossings and river operations; reduced off-road maneuver capacity; increased maintenance costs; increased flood control/erosion prevention measures; transportation infrastructure damage Degradation or loss of coastal areas and infrastructure; increased cost of infrastructure reinforcement to withstand increased storm intensities; increased cost of infrastructure modification (e.g., raising pier heights); impacts to littoral and shore training and ranges; increased regulatory constraints on training land access; impacts on supply chain from potential shipping interruptions; increased demand for freshwater resources and associated increased cost of saltwater intrusion countermeasures; impact to future land availability and siting of new construction
Changes in ocean temperature, circulation, salinity, and acidity	Potential greater change to global climate system; negative impacts to general populations that rely upon fish as their main source of protein; coral reef losses that may	Exacerbation of conditions and mission impacts discussed above; coastal installation vulnerability; regional instability; increased potential

Climate Change Phenomena	Potential Impacts	Potential Mission Vulnerabilities
	impact ocean productivity and storm surge/wave dampening benefits	for conflict or humanitarian assistance

3. Process for DoD Adaptation Planning and Evaluation

The QDR provides broad direction for future DoD strategies that will define plans and policies. Prompted by the QDR, the Department is prudently considering how to factor climate impacts into its mission areas. Given the diversity and complexity of DoD's mission and operations, there is an equally wide array and magnitude of planning processes across DoD. The Department recognizes that both operational and infrastructure plans and processes present opportunities to integrate climate change risks and opportunities to enhance the resilience of our mission, at home and abroad.

DoD is well-versed in employing systematic methodologies and modeling frameworks in order to assess potential threats and risks to national security. The use of these risk assessment tools is an essential element of accomplishing the DoD mission. The Department anticipates employing a similar risk-based approach to evaluate multiple scenarios of potential climate change effects on the DoD mission. Many of the Department's current efforts are focused on assessing potential climate change impacts to, and adaptation strategies for, facilities, built infrastructure, key ecosystems and protected species, and capabilities where military training is conducted or supported, and evaluating potential actions DoD can take to respond to these impacts. Sections 4 and 5 discuss specific efforts.

DoD intends to move forward with the previously stated goals for adaptation planning and evaluation. Goal implementation is described below.

> "Preventing wars is as important as winning them, and far less costly." (p. 7)
>
> — 2011 National Military Strategy.

Goal 1. Define a Coordinating Body to Address Climate Change

The Department intends to define an appropriate structure utilizing existing bodies and organizations within DoD to guide the development, implementation, and evaluation of climate-related policy, guidance, and practice. The Senior Sustainability Council (SSC) is currently responsible for coordinating climate change adaptation efforts. The SSC will establish a technical advisory committee or working group to take direction from and provide advice to the council regarding the state of climate science, vulnerability and impact assessment, and adaptation science and practice. The advisory committee will analyze technical constraints and considerations related to climate change-related policy, guidance, and practice. The advisory committee will focus on ensuring that the Department has access to the climate-related information necessary to make informed decisions that support the Department's mission. This structure would identify those offices and existing forums, with authority in this area, those that would assist in coordination and guidance, and those that would be involved in support and implementation.

> The DSB recommended the Secretaries and Chiefs of the Services should: better integrate climate change and disaster risk reduction consideration into exercise, training, and educational materials; establish metrics focused on risk reduction to minimize the impact of climate change on military and support operations, forces, programs, and facilities; ensure climate change resilience by incorporating climate change risk in design standards for facilities and installations, with an emphasis related to energy- and water-intensive uses.
>
> — Defense Science Board "Trends and Implications of Climate Change for National and International Security," Oct 2011.

The advisory committee, once established, will:

- Optimize use of existing plans and processes and identify gaps where new policies could be developed;
- Stress the importance of the science-policy interface;
- Foster sound vulnerability and impact assessment;

- Emphasize iterative and adaptive policy and planning approaches; and
- Monitor assessment and adaptation implementation effectiveness, learn from these experiences, and adjust action when needed.

Goal 2. Utilize a Robust Decision Making Approach Based on the Best Available Science

Assessing climate change vulnerabilities, impacts, and adaptive responses requires a deliberative and iterative approach. The Department intends to develop appropriate assessment tools for use across all affected DoD Components. In developing its approach to assessment, adaptation planning, and implementation, the Department will strive to:

- Establish a process to obtain updated scientific data on potential future climate conditions and potential impacts;
- Use commonly accepted future climate scenarios that are based on the best available science, recognize uncertainties, and updated as the science changes;
- Provide guidance so that assessments consistently apply science that is appropriate in terms of location, resolution, and timeframe; and
- Use pilot approaches to develop decision frameworks for assessment and adaptation planning that attempt to match decisions to available and appropriately down-scaled climate information and other data.

Goal 3. Integrate Climate Change Considerations into Existing Processes

Climate change and climate variability will affect many of the Department's activities and decisions related to future operating environments, military readiness, stationing, environmental compliance and stewardship, and infrastructure planning and maintenance. Climate change also will interact with other stressors that the Department now considers and manages. As a result, adaptation to climate change and variability should not be a separate decision-making process, but rather an aspect of overall management. DoD intends to fully integrate climate change considerations into its extant policies, planning, practices, and programs. Some stand-alone policy and guidance may be needed to help direct specific assessment activities and adaptation implementation; however, by and large the Department will use existing mechanisms to implement policy and guidance and to ensure mission and environmental sustainability.

Goal 4. Partner with Federal Agencies and Allies on the Challenges of Climate Change

Partnerships will be needed to fully ensure DoD's mission is sustainable under climate change. The Department cannot assess its vulnerabilities and implement adaptive responses at its installations if its neighbors and stakeholders are not part of the process. Decisions made by outside communities will affect DoD and DoD's decisions will also affect outside communities. Moreover, aspects of our mission such as force deployment may be affected by assets outside DoD control, such as transportation infrastructure.

> "In combination with U.S. diplomatic and development efforts, we will leverage our convening power to foster regional and international cooperation in addressing transnational security challenges." (p. 15)
>
> — 2011 National Military Strategy.

The requisite scientific and practical understanding needs to be obtained in concert with the rest of the Federal community. This can occur through partnerships with individual agencies such as the National Oceanic and Atmospheric Administration or through the Department's continued participation in forums such as the National Climate Assessment and informal forums such as the Interagency Forum on Climate Change Impacts and Adaptations.

Internationally, the Department will continue its collaboration with the State Department and foreign militaries on vulnerability assessment and adaptation efforts. The Department has already started to assess potential climate change impacts and begin initial adaptation planning. Efforts to partner with foreign defense force counterparts are coordinated through existing planning processes. Climate change presents a unique opportunity to work collaboratively in multilateral forums, promoting a balanced approach that will improve human and environmental security in the region. The Department's disaster response programs will continue to provide domestic and international response, but should adapt its response planning based on plausible climate change scenarios.

4. Actions to Better Understand Climate Change Risks and Opportunities

DoD is already working to foster efforts to assess, adapt to, and mitigate the impacts of climate change. The Military Services are considering potential climate change vulnerabilities and impacts to their activities and infrastructure in light of their Service-specific missions and plans.

> "Our diplomacy and development capabilities must help prevent conflict, strengthen weak and failing states, lift people out of poverty, combat climate change and epidemic disease ..." (p. 11)
>
> — 2010 National Security Strategy.

The Department looks to the Strategic Environmental Research and Development Program (SERDP), a joint effort among DoD, the Department of Energy, and the Environmental Protection Agency, to develop climate change assessment tools for DoD's installations. The DoD Legacy program can be used in transitioning these tools for natural and cultural resources management applications. The Navy's Arctic and Climate Change Roadmaps also outline specific action items which contribute to DoD's understanding of how a changing climate can pose risks and opportunities to its mission and operations. The Air Force 2010-2030 Strategic Environmental Assessment includes discussion of climate change as a strategic consideration for Air Force strategic planners. The Army is investigating climate risks to installation lands and facilities in its Environmental Quality Technology research program, and the Army Climate Change Workgroup is developing a framework for integrating climate change considerations into existing planning processes. The Sustainable Ranges Integrated Product Team, led by an Office of the Secretary of Defense and tasked to address test and training encroachment and sustainability issues, also includes consideration of climate change as an emerging encroachment issue. As discussed earlier in this roadmap, such nascent DoD initiatives will benefit significantly when an overarching DoD policy framework can be put in place to help guide and focus such efforts.

The sections that follow summarize activities currently underway to understand the risks and opportunities to DoD operations. Some of these assessments are general and high-level, while others are specific to certain subject matter areas and/or locations.

> "Our diplomacy and development capabilities must help prevent conflict, strengthen weak and failing states, lift people out of poverty, combat climate change and epidemic disease ..." (p. 11)
>
> — 2010 National Security Strategy.

4A. General Assessments

DoD is working to overlay regional climate models with installation locations, in order to appropriately downscale climate variables for individual locations and develop an analytical tool that can be used to generate climate projections at the regional level. DoD is involved in high-level climate and weather data gathering efforts, as the Air Force 14th Weather Squadron collects, stores, and characterizes earth-space environmental data, receiving nearly 500,000 weather observations and satellite-derived wind profiles each day and sharing these data with the National Climatic Data Center and the Navy's Fleet Numerical Meteorological and Oceanographic Detachment. DoD collaborates with the National Oceanic and Atmospheric Administration on the development and operational implementation of a national Earth System Prediction Capability.

4.B. Coastal Risks and Opportunities

Many of DoD's military installations are concentrated in coastal regions of the continental United States. As a result, DoD is undertaking multiple projects to assess climate change impacts to these installations and areas. Several of these projects focus specifically on sea level rise and storm surge, developing the necessary methodologies and/or tools that might inform decision making processes, including where to build and how to update coastal installations. Other projects deal with climate impacts on coastal ecosystems, as the military's long-term use of coastal installations is, in part, dependent on the ability to maintain the continued functioning of coastal ecosystems. Projects that specifically address coastal ecosystems can help educate natural resource managers and enhance their decision making processes related to managing these ecosystems for their training/testing value, storm protective functions, and species diversity. The Department, drawing on the lessons learned from the preceding studies, has identified the key technical considerations to consider when conducting assessments of climate change impacts on coastal military installations. This effort will assist the Department in developing its approach to coastal assessment.

4.C. Arctic Risks and Opportunities

The Department has also begun to assess and plan for changes to our operating environment. For example, preliminary assessments have been conducted for the Arctic where measurable climate change impacts are already occurring. These efforts have focused on assessing the Department's Arctic observing, mapping, and environmental prediction capabilities, as well as identifying science and technology needs. The Department has completed two Capabilities Based Assessments for Arctic surface and environmental prediction capabilities and a Fleet Readiness Assessment. The Department is developing cooperative partnerships with interagency and international Arctic stakeholders to collaboratively address future opportunities and potential challenges inherent in the projected opening of the Arctic.

4.D. Permafrost in Alaska

The change in permafrost in Alaska is impacting both the built and natural infrastructure. The Department held, as early as 2009, workshops to better understand affected defense assets and military missions in Alaska. The melting permafrost will impact foundations, utilities, runways and roads. This is a challenge for operation and maintenance especially when considering 80% of the infrastructure that will exist in 2050 is already in place today. The melting permafrost influences on training lands and natural ecosystems can significantly affect the types and timing of training activities. The potential ecosystem responses in interior Alaska to climate change could have severe ramifications on how, where, and when the DoD can train in Alaska. To address concerns related to climate change's impact on permafrost freeze and thaw processes and other ecological factors in interior Alaska, DoD initiated a suite of projects in FY 2011 focused on understanding and predicting these changes and the implications for Alaskan training land sustainability. These efforts will fill knowledge gaps relative to how climate change is affecting permafrost and the overall system dynamics, informing decisions on the development of future training and installation management plans.

4.E. Arid Ecosystems

Long-term use of military installations and ranges in the southwestern United States depends, in part, on the condition of local ecosystems. Changes to local ecosystems can adversely impact natural resources and affect the use of certain locations for training, and/or increase the possibility of wildfires. DoD has initiated several projects to assess changes to ecosystems in the southwestern United States, including the intermittent and ephemeral stream systems that harbor much of the region's biological diversity, and the interaction of land-use activity, altered water sources, the introduction of invasive species, and altered fire regimes.

4.F. Pacific Islands

In FY 2013, DoD anticipates initiating climate change studies to assess the impacts on DoD facilities in the Pacific. Changes in sea level, precipitation, and storm patterns can have significant impact on the island infrastructure that supports DoD missions in the region.

5. Actions to Address Climate Change Risks and Opportunities

In addition to the activities outlined in Section 3, DoD's current efforts to integrate, partner, and undertake pilot activities to address climate change risks and opportunities include the following.

DoD is already beginning to incorporate climate considerations into installation-level planning, as well as training plans. The Department is starting to incorporate climate change science and strategic considerations into formal training and education. The Military Services are beginning to explore incorporating climate risk/vulnerability factors into installation development planning processes. At the DoD level, United Facilities Criteria (UFC) 2-100-01, paragraph 3-5.6.2.3 requires master planners to consider climatic changes (including but not limited to: changes in land use and population density in the vicinity of installations; changes in climatic conditions such as temperature, rainfall patterns, storm frequency and intensity and water levels) when crafting long-range installation infrastructure master plans. UFC 2-100-01, paragraph 3-5.6.2.3 specifically calls out the National Climate Assessment as a source for reliable and authorized climate change scenarios. The Department's Natural Resources Conservation Program Instruction (DoDI 4715.03) requires installation natural resources management plans (INRMP) to assess the potential impacts of climate change on natural resources and to adaptively manage such resources to minimize adverse mission impacts.

> "We must, therefore...design structures and systems that can withstand disruptions and mitigate associated consequences, ensure redundant systems where necessary to maintain the ability to operate, decentralize critical operations to reduce our vulnerability to single points of disruption, develop and test continuity plans to ensure the ability to restore critical capabilities, and invest in improvements and maintenance of existing infrastructure." (p. 27)
>
> — 2010 National Security Strategy.

As part of its Sustainable Ranges Initiative, DoD has conducted research and completed an initial study of potential climate change vulnerabilities affecting DoD training and potential adaptive measures. Additional research and coordination is ongoing, and several workshops have been held to engage with DoD offices and Federal agencies on possible avenues to foster a more adaptive individual and organizational culture that is better prepared to respond to mission stressors such as climate change. DoD is also actively engaged with regional partnerships in the Southeastern and the Southwestern U.S. Both regions are very significant to DoD, and host a number of major military installations and ranges. The aim of both the Southeast Regional Partnership for Planning and Sustainability and the Western Regional Partnership is to strengthen regional coordination and advance the missions and land use objectives of

DoD and the other state and Federal agencies involved. Both partnerships are actively assessing the climate change challenge, along with a number of other often interrelated issues (habitat and species protection, land use planning, energy development, coastal zone management, fire management and disaster preparedness, and sustainable land use) as they work on cooperative policy and planning initiatives. DoD expects cooperation on climate change issues to continue and likely grow in importance within both partnerships in coming years.

Through SERDP, DoD has initiated pilot projects intended to develop and test assessment approaches and decision-making frameworks for climate adaptation appropriate for military installations. These pilot efforts will help DoD identify appropriate processes for matching climate information with DoD decision processes, understanding data needs for vulnerability assessments, and developing adaptation tools with installations across the country.

INDEX

A

B

C

D

E

F

N

O

P

R

W